Latino
WOMEN
OF
SCIENCE

by Leonard Bernstein
Alan Winkler
Linda Zierdt-Warshaw

Peoples Publishing Group, Inc.

Free to Learn, to Grow, to Change

1-800-822-1080

Project Manager, *Doreen Smith*
Copy Editor, *Sal Allocco*
Cover Design, *Jeremy Mayes, Doreen Smith*
Design, *Kristine Liebman, Doreen Smith*
Illustrations, *Jeremy Mayes, Brooke Kaska*
Production/Electronic Design, *Brooke Kaska, Janet Kliesch*
Photo Research, *Brooke Kaska*

PHOTO CREDITS

p.4, Margarita Colmenares; p.8, Jane Delgado; p.12, Irma Gigli; p.16, Ana Sol Gutierrez; p.20, Romy Ledesma; p.24, Aliza Lifshitz; p.28, Lissa Martinez, special thanks to Maureen Ruivo; p.32, Harvard Observatory; p. 36, The Bancroft Library, University of California, Berkeley; p.40, Antonia Novello; p.44 Adriana Ocampo; p.48, NASA; p.52, Steve Covallo; p.56, Sandy Nolte, American Public Health Association; p.60, Carol Sanchez; Clip Art from CorelDraw Corp.

ISBN 1-56256-705-5

Peoples Publishing Group, Inc.
299 Market Street
Saddlebrook, NJ 07663

Printed in the United States of America.

10 9 8 7 6 5 4

CONTENTS

Each chapter contains a biographical sketch of a notable Latino woman of science, an activity called It's Your Turn, and a Think Work Act page with questions and activities featuring a variety of skills and learning techniques.

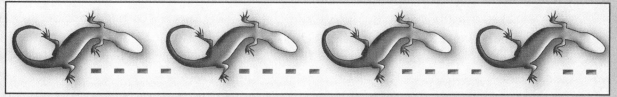

MARGARITA COLMENARES

1957 -

1957 Margarita Colmenares is born in Sacramento, California

1989 Colmenares becomes the national president of the Society of Hispanic Professional Engineers

1994 Colmenares is appointed as Director of Corporate Liaison by Secretary of Education Richard Riley.

1981 Colmenares receives her B.S. in civil engineering from Stanford University.

1991 Colmenares is selected to participate in the White House Fellowship Program.

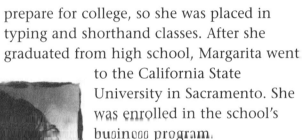

The White House Fellowship Program was started in 1964. The program provides young people with an opportunity to become involved in the work done by the federal government. People chosen for the program go to Washington, D.C., and work within the government for one year. In 1991, Margarita H. Colmenares, a civil engineer from California, joined the program. She became the first Latin American engineer to do so.

Margarita Colmenares was born in Sacramento, California in 1957. She was the oldest of five children. Margarita's parents were immigrants who came to the United States from Oaxaca, Mexico.

Margarita's parents valued education. Although they did not make much money, the Colmenares' provided all their daughters with a private Catholic school education.

Colmenares went to an all-girl high school. At the time, girls were not encouraged to prepare for college, so she was placed in typing and shorthand classes. After she graduated from high school, Margarita went to the California State University in Sacramento. She was enrolled in the school's business program.

During her first year in college, Colmenares discovered that her interests were not in business but in engineering. However, she did not have the educational background needed to study engineering. To get the background, Colmenares began to take courses in science and math at the Sacramento City College. She also took a part-time job with the California Department of Water Resources. In this job, Colmenares inspected dams and water-purification plants. She was looking for signs of cracks and other damage. The job helped Colmenares pay for her education. It also gave her hands-on experience in engineering.

Using money earned from her job and from scholarships, Colmenares continued her engineering studies at Stanford University.

Margarita Colmenares

While there, she also took part in community activities. One such activity was as an instructor and a co-director of the Stanford Ballet Folklorico. The group performed at local fairs and in the homes of senior citizens.

Before her last year at Stanford began, Colmenares took part in the Chevron Corporation's Co-Op Education Program. For nine months, she worked full-time for the company.

In 1981, Colmenares graduated from Stanford University with a B.S. in civil engineering. Following graduation, she began to work for Chevron as a field construction engineer. In this job, Colmenares worked closely with the public.

Colmenares spent more than 10 years with the Chevron Corporation. While with the company, she held many positions. In 1983, she became a foreign training representative. In this job, she worked with engineers from around the world.

Colmenares later went to Texas to become a compliance specialist for her company. This job involved making sure that the company met environmental, fire, health, and safety laws.

Colmenares has been active in many organizations. In 1982, she founded the San Francisco chapter of the Society of Hispanic Professional Engineers (SHPE). She was chapter president until 1984. In 1989, Colmenares became the national president of SHPE. To make time for this role, she took a leave of absence from her job.

Colmenares has also been active in community work. In the 1980s, she often performed with a Mexican folk-dance group. While working in Houston, Colmenares was a panelist for the Cultural Arts Council. She was also a member of the board of directors of the Texas Hispanic Women's Network.

In 1991, Colmenares was selected as one of 16 people to take part in the White House Fellowship Program. Her selection into the program was

> **VOCABULARY**
>
> A CIVIL ENGINEER is trained in the design and building of public works.
>
> A LIAISON works to coordinate activities or to bring people together to work on a common goal.

made largely because of her leadership ability and her commitment to community service. To take part in the program, Colmenares was granted a second leave of absence from her job.

While in the White House Fellowship Program, Colmenares worked in the Department of Education. She worked as a special assistant to Deputy Secretary of Education, Madeline Kunin. In this job, Colmenares looked for ways to improve math and science education in U.S. schools. As part of her research, she visited schools in other countries to meet with educators and observe their teaching methods.

In 1992, Colmenares returned to Chevron. She remained with the company for two more years. During this time, Colmenares worked as a project manager.

In 1994, Colmenares returned to the Department of Education. She was appointed Director of Corporate Liaison. The appointment was made by Secretary of Education Richard Riley. In this job, Colmenares looks for ways to involve U.S. businesses in improving education.

Colmenares has received many honors for her civic and professional activities. In 1989, she was given the Community Service Award by *Hispanic Engineer* magazine. She was also named SHPEs Role Model of the Year. In 1990 and again in 1992, *Hispanic Business* magazine chose Colmenares as one of the 100 most influential Hispanics in the United States. In 1991, she was given the Pioneer Award by the National Hispanic Engineers Achievement Awards Conference.

In 1992 and 1993, Colmenares was named to the "All-Star Team" of nationally recognized leaders in engineering. The awards were presented during National Engineer's Week. One reason Colmenares received the award was because of the time she spends encouraging young people, especially Latinos, to seek careers in the fields of science and engineering.

It's YOURTURN

Hands-On Activity

CLEANING DIRTY WATER

MATERIALS (per 2 students)

sand, clay, and water mixture saturated alum solution, top third of plastic beverage container, coffee filter, 2 large plastic cups; teaspoon, sink or basin, clock

SAFETY

Wear safety goggles and a laboratory coat or apron throughout this activity. Clean up any spills that occur immediately.

BACKGROUND INFORMATION

Clean drinking water is necessary to human health. Water-purification plants work to make water clean. To do this, these plants make use of filtration, sedimentation, and coagulation. Filtration involves passing water through a material that allows certain matter to pass through while trapping other matter. During this process, even some tiny organisms are removed from the water. During sedimentation, water is permitted to stand until some heavy materials in the water settle out. During coagulation, a chemical is added to water. The chemical causes small particles in the water to form heavy clumps. These clumps then settle out of the water.

PROCEDURE

1. Half-fill one cup with the sand, clay, and water mixture. Use the spoon to stir the mixture. Observe and describe the appearance of the mixture.

2. Turn the beverage container upside-down and place it into an empty cup. Place a coffee filter into the neck of the container.

3. Carefully, pour the mixture into the beverage container. Observe and describe the material collected in the cup below the container.

4. Set the cup with the liquid aside, where it will remain undisturbed for 10 minutes. After 10 minutes, observe and describe the appearance of the mixture.

5. Add 2 spoonfuls of the alum solution to the cup. Use the spoon to gently mix the alum with the liquid part of the cup's contents. Try not to disturb matter at the bottom of the cup.

6. Set the cup aside and allow it to remain undisturbed for 10 minutes. Observe and describe the appearance of the mixture.

ANALYZE AND CONCLUDE

Answer the questions on the lines provided.

1. What role did gravity play in the purification of the water?

_____

_____

2. Which cleaning method seemed to remove the most impurities from the water?

_____

3. What materials other than a coffee filter might be use in the filtration process?

_____

_____

4. Fish and frogs may live in the body of water from which water is obtained. How might these animals affect water quality?

_____

_____

   Margarita Colmenares

Name _____ Date _____

Think WORK ACT

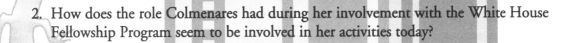

CRITICAL THINKING *Answer the following questions in complete sentences.*

1. What problems did Margarita Colmenares have to overcome to pursue a career in engineering?

_____

_____

2. How does the role Colmenares had during her involvement with the White House Fellowship Program seem to be involved in her activities today?

_____

_____

3. What types of problems may result from cracks and other damage to dams and water purification plants?

_____

_____

GOING FURTHER   *Complete three of the following.*

USING COMMUNITY RESOURCES

Visit or write to several colleges to find out what courses are needed to prepare for a career in engineering. Use the information you obtain to outline a course of study for a high school student who wants to pursue a career in engineering.

COOPERATIVE LEARNING

Margarita Colmenares is very involved in community activities. Prepare and carry out a survey to find out in what community activities students in your school are involved. Have each group member survey 10 people. Combine all the results into a group summary.

CONCEPT MAPPING

Use reference books to research the steps involved in the water purification process. Create a flowchart that outlines and describes each step.

RESEARCH AND REPORT

Interview an engineer who works for a company in your area. Prepare at least ten questions to ask this person about the training and the duties their job involves. During the interview record the questions and the responses you get. Prepare a report summarizing the interview.

JOURNAL WRITING

Write about the career you think you would most like to pursue when you finish school. Describe how the courses you are currently taking might be helpful in pursuing this career. Identify what other courses (both high school and college) you think you might need to take.

JANE DELGADO

1953 -

1953 Jane Delgado is born in Havana, Cuba.

1972 Delgado graduates from S.U.N.Y. with a B.A. in psychology.

1979 Delgado begins work with the U.S. Department of Health, Education, and Welfare.

1985 Delgado is appointed president and CEO of the Coalition of Spanish Speaking Mental Health Organizations.

1969 Delgado enters college at age 16.

1975 Delgado receives her M.A. in psychology.

1981 Delgado receives her Ph.D. in clinical psychology and an M.S. in urban policy and sciences.

Health-care issues concern everyone. Such issues include the kinds of health services needed, who is served, costs, and treatment. Managing all these areas is a huge task. But this was the job of Latin American psychologist Dr. Jane L. Delgado, when she began work for the Secretary of Health and Human Services (HHS).

Jane Delgado was born in Havana, Cuba, in 1953. Her father, Juan Lorrenzo Delgado Borges, was a magazine publisher. Her mother, Lucila Aurora Navarro, took care of the family.

When Jane was two, the family moved to New York City. After the move, Jane's father left the family. Her mother, who spoke little English, took a low-paying factory job to support and make a stable home for her daughters. Even with her low wages, she was able to save money for Jane's college education.

When Jane entered school, she spoke no English. However, she enjoyed school and learned quickly. By the time she reached third grade, she had mastered English and was placed in a class for gifted children.

Jane did well in school throughout her elementary and junior high school years. However, when she entered high school, she was no longer challenged by her studies. Jane became bored with school. Eager to graduate, she designed a program that allowed her to graduate early. On her own, Jane researched colleges and filed her applications. In 1969, at the age of only 16, Jane entered the State University of New York (S.U.N.Y.) at New Paltz. Although younger than her classmates, Jane worked hard at her studies and became involved in school activities. Only three weeks after starting college, she was elected vice-president of her freshman class.

In 1972, Jane Delgado graduated from S.U.N.Y. with a B.A. in psychology. She was only 19 at the time. Delgado then moved back to New York City.

After returning to New York City, Jane began work with the Children's Television Network. She soon became the talent coordinator for the children's educational program, *Sesame Street*. In this role, Jane selected the children who would appear on the program. While working in this job, Jane also began work on a master's degree in psychology. She obtained her M.A. in psychology from New York University in 1975.

Two years after receiving her M.A., Delgado was accepted into the Ph.D. program of S.U.N.Y. at Stony Brook. While working on her degree, Delgado also worked at the school as a teaching assistant. When not working at the university, she taught courses at the New York Experimental and Bilingual Institute.

Jane Delgado received her Ph.D. in clinical psychology in 1981. The same year, she also earned a second master's degree in urban policy and sciences. This degree was received from the W. Averell Harriman School of Urban and Policy Studies. This second master's degree helped Delgado in her work as a health-care administrator.

Delgado began work in health-care administration in 1979 with the U.S. Department of Health, Education, and Welfare (HEW). At the time, she was still working on her Ph.D. In her job with HEW, Delgado managed projects dealing with issues of Latin Americans, African American colleges, and undocumented workers.

In 1980, the name of the Department of Health, Education, and Welfare was changed to the Department of Health and Human Services (HHS). Through its secretary, HHS advises the president on matters concerning the health and welfare of U.S. residents. The Office of Community Services (OCS) is a part of HHS. Delgado worked for this agency and served as its chief. Part of her job with OCS was to manage

> **VOCABULARY**
>
> PSYCHOLOGY is the branch of science that studies behavior and mental processes

$400 million in grants to states. Delgado ran the program and reported to congress about how funds should be spent.

Delgado's work earned her a post in the office of Margaret Heckler, the Secretary of HHS. Here, Delgado supervised health- and service-related programs. She advised Secretary Heckler about minority health-care issues. Delgado also acted as the senior policy coordinator for national committees dealing with mental health and drug and alcohol abuse.

In 1985, Delgado was appointed president and CEO of the Coalition of Spanish Speaking Mental Health Organizations (COSSMHO). The next year, this group was renamed the National Coalition of Hispanic Health and Human Services Organizations. The role of the agency is to improve health and human services for Latin Americans.

In her first year with COSSMHO, Dr. Delgado began a program to help educate Latin Americans about AIDS. She has also tried to increase knowledge about health-care issues related to women and the environment. In addition, COSSMHO often provides health information about Latin Americans to the U.S. government.

In addition to her work with COSSMHO, Delgado also chairs the National Health Council (NHC). The NHC is made up of many health organizations. Two of the most well known are the American Cancer Society and the American Heart Association. Delgado also serves on the boards of many health councils, where she tries to make sure the needs of Latin Americans are addressed. She is currently publishing her book *Salud! A Latina's Guide to Total Health, Body, Mind and Spirit.*

It's **YOUR TURN**

ANALYZING DRUG ABUSE DEATH RATES

BACKGROUND INFORMATION

Drugs are used to treat disease and to reduce pain. When used properly, drugs save lives. When used improperly, drugs become poisons that destroy lives through addiction, crime, and death. Agencies within the Department of Health and Human Services are responsible for administering programs to prevent and control drug abuse. These agencies include the Public Health Service and the Substance Abuse and Mental Health Services Administration.

PROCEDURE

1. Study the table. Use the data in the table to answer the questions in the Analyze and Conclude.

Death Rates Resulting from Drug Abuse in the U.S.	
Year	Deaths (per 100,000)
1980	6,900
1985	8,663
1990	9,463
1991	10,388
1992	11,703

ANALYZE AND CONCLUDE

Answer the following questions on the lines provided.

1. What does the data in the table show?

   _____

   _____

2. From the data provided, what can you conclude about the number of deaths resulting from drug abuse per 100,000 people since 1980?

   _____

   _____

3. How many deaths per 100,000 people occurred in 1990?

   _____

4. In which year did the least number of deaths per 100,000 people occur? The greatest number?

   _____

5. Based upon the trend shown in the table, would you expect the reported number of drug-related deaths to increase or decrease by the year 2000? Explain.

   _____

   _____

   _____

Think WORK ACT

CRITICAL THINKING *Answer the following questions in complete sentences.*

1. How might a degree in psychology be helpful to a person who studies health issues such as drug abuse and alcoholism?

   _____

   _____

2. What do you think the study of urban policy involves?

   _____

   _____

3. What types of health-care issues might be unique to minorities?

   _____

   _____

GOING FURTHER

Complete three of the following.

BUILD YOUR PORTFOLIO

Write a brief autobiography that could be used to help you prepare for admission to a college. Be sure to include information about the field you hope to study and why.

COOPERATIVE LEARNING

Find information about health-care issues of women and minorities. Organize a health fair to make people in your community more aware of these issues. Contact health agencies to obtain brochures that can be made available to people attending the fair. Also arrange for speakers for the event.

CONCEPT MAPPING

Do library or computer research to find out what agencies are part of the U.S. Department of Health and Human Services. Organize your findings in a concept map.

RESEARCH AND REPORT

Do research to find out what kind of work is done by the Centers for Disease Control (CDC) and the National Institutes of Health (NIH). Find out how this work compares to that done by the World Health Organization (WHO). Write a report of your findings.

USING COMMUNITY RESOURCES

Use a phone directory to identify five community organizations that deal with health care. List the names, addresses, and telephone numbers of the organizations. Call or write to each agency to find out what services or information it provides to the public. Summarize your findings in a table.

IRMA GIGLI

1936 -

1936 *Irma Gigli* is born in Cordoba, Argentina.

1961 *Gigli* receives her certification in dermatology in New York.

1967 *Gigli* returns to the U.S. and serves as chief of dermatology at Harvard Medical School.

1982 *Gigli* becomes head of the dermatology department at the University of California at San Diego.

Today *Gigli* is chief of dermatology at the Universidad Nacional de Cordoba in Argentina.

1958 *Gigli* moves to the United States and completes a residency in dermatology at the Cook County Hospital, Chicago.

1964 *Gigli* makes a brief return to Argentina. She then travels to Germany to study at the University of Frankfurt.

1974 *Gigli* goes to England to study with Nobel Prize-winner, Rodney Porter.

1985 *Gigli* marries immunochemist Hans Müller-Eberhard.

The department head at a college is a respected post. Such posts are usually held by men; the appointment of a woman is rare. It is even more rare for a woman to head a department in a medical field. However, Latina physician Irma Gigli has held such posts. Gigli was the former head of the dermatology department at the University of California-San Diego, School of Medicine. She is one of only a few women in the world to have held such a position.

Irma Gigli was born in Cordoba, Argentina, in 1936. She was educated, from elementary school through medical school, in her homeland. In 1958, Gigli came to the United States on a student visa. When she arrived, she already had both a teaching degree and a degree in medicine.

After coming to the United States, Gigli went to Chicago, Illinois. Here she completed an internship and a residency in dermatology at the Cook County Hospital. Dermatology is the branch of medicine that deals with the health of the skin. Gigli completed her studies at the Cook County Hospital in two years.

In 1960, Gigli moved to New York. She studied at the New York University's School of Medicine. The following year, she received her certification in dermatology. Gigli then moved to Miami, Florida, to attend the Howard Hughes Medical Institute.

In 1964, Gigli's student visa expired. She had to leave the United States and return to Argentina. When she arrived, Argentina was in a state of political unrest. On the advice of her family, Gigli again left Argentina. This time, she traveled to Germany. In Germany, Gigli worked as a scientist in the department of dermatology at the University of Frankfurt. She remained in Germany for two years.

In 1967, Gigli returned to the United States. She went to Massachusetts. She later joined the faculty of the Harvard University Medical School, having obtained one of the early Howard Hughes Medical Institute Awards. At Harvard, Gigli did research in immunology, the medical field that studies the body's resistance to disease. Gigli also taught dermatology, served as the chief of dermatology, and worked at several Harvard University hospitals.

In 1974, Gigli traveled to England. She made the trip to study with Nobel prize-winning scientist Rodney Porter at Oxford University. While in England, Porter offered Gigli a position at Oxford. However, Gigli did not want to again make her home in a new country. She turned down the offer and returned to the United States in 1976.

After returning to the United States, Gigli took a job at New York University. She worked as a professor of dermatology and experimental medicine. Gigli held this post for six years. She then went to work for the University of California at San Diego. Here, Gigli worked as a professor of medicine and headed the dermatology department. Gigli also served as an associate chair for research in the university's department of medicine.

In 1985, Gigli married Hans Müller-Eberhard, an immunochemist. When they married, Müller-Eberhard worked for the Scripps Clinic in La Jolla, California. Three years later, he became director of the Bernhard Nocht Institute for Tropical Medicine. The institute is located in Hamburg, Germany. Through her husband, Gigli also rekindled an interest in tropical diseases.

Irma Gigli's interest in tropical diseases continues today. Each year, she does research at the institute where her husband works. She also does much research on autoimmune diseases, such as AIDS and lupus. In 1995, Gigli and her husband joined the University of Texas, Houston, to develop the Institute of Medicine for the Prevention of Humane Diseases, a new Institute within the University of Texas Medical Center.

Irma Gigli has published hundreds of papers about immunology and tropical diseases. She has also been admitted as a member of the most prestigious organizations in medicine, including the Institute of Medicine of the National Academy of Sciences, the American Association of Physicians, and the American Society for Clinical Investigation. Gigli's membership in these groups is a great accomplishment because the groups admit members only by invitation.

In addition to her work as a teacher and researcher, Irma Gigli is also well known for her public speaking. She has lectured in the United States, in Central and South America, and in Europe.

VOCABULARY

A VISA is a permit that allows a person to travel in a certain foreign country for a specific period of time.

An IMMUNOCHEMIST is a scientist who studies the chemical factors involved with the body's resistance to disease.

An AUTOIMMUNE DISEASE results when the body attacks its own cells as if they were agents of disease.

It's YOUR TURN

TROPICAL DISEASES AND THEIR TRANSMISSION

MATERIALS (per student)

pencil, access to a reference library

BACKGROUND INFORMATION

The *tropics* is a geographic zone located between the Tropic of Cancer and the Tropic of Capricorn. Tropical areas are typically hot and humid. In addition, rainfall may average from 60 to 100 inches per year. Heat, humidity, and rainfall provide ideal growth conditions for many living things, especially insects and microorganisms. The diseases that occur in the tropics are often referred to as tropical diseases. Examples of such diseases include malaria, yellow fever, sleeping sickness, elephantiasis, and yaws.

PROCEDURE

1. Use biology texts and other references to fill in the information that is missing from the table.
2. Use your completed table to answer the Analyze and Conclude questions.

Tropical Diseases and Their Transmission			
Disease	*Agent of Disease*	*How Transmitted*	*Where Disease Occurs*
Malaria			
Yellow Fever			
Sleeping Sickness			
Elephantiasis			
Yaws			

ANALYZE AND CONCLUDE

Answer the following questions on the lines provided.

1. What agent is responsible for elephantiasis?

_____

2. Which disease is caused by a virus?

_____

3. What do malaria, yellow fever, and some forms of elephantiasis have in common?

_____

_____

4. In which part of the world do most of the tropical diseases listed occur?

_____

5. Which disease is not transmitted by an insect?

_____

6. Why is mosquito control important to help reduce the number of cases of certain tropical diseases?

_____

_____

Think WORK ACT

CRITICAL THINKING *Answer the following questions in complete sentences.*

1. How do you think the concerns of dermatology and immunology are related?

   _____

   _____

2. Malaria is sometimes called swamp fever. Using this information, what can you infer about the types of areas in which malaria is common?

   _____

   _____

3. What are some precautions you might take to prevent being bitten by insects that carry disease-causing organisms?

   _____

   _____

GOING FURTHER

Complete three of the following.

COOPERATIVE LEARNING

Add to the table in the activity by doing research to identify five other tropical diseases, the agent that causes each disease, how each disease is transmitted, its treatment, and where it is common. Combine each group member's findings in a table and on a world map.

RESEARCH AND REPORT

One role of the Centers for Disease Control (CDC) is to monitor the spread of infectious diseases. Although the agency's chief concern is maintaining the health of Americans, its doctors often travel to remote areas to study outbreaks of new diseases. Do research to find out what recent outbreaks of disease the CDC has been researching. Write a report of your findings including where in the world each disease has been observed.

CONCEPT MAPPING

Do research to find out what role the skin plays in protecting the body from disease. Also find out what other parts of the body are involved in fighting disease. Develop a concept map to summarize your findings.

PERFECT YOUR SKILL

Use a biology text or library resources to learn about the life cycle of the parasite that causes malaria and how it causes the disease. Develop an outline or drawing that traces the steps involved in the spread of malaria.

JOURNAL WRITING

Describe some of the things you do to keep your skin healthy. Explain what benefit you think each practice has to your skin.

ANA SOL GUTIÉRREZ

1942 -

1942 **Ana Sol** is born Ana Sol in El Salvador.

1976 After having earned her master's degree, **Gutiérrez** is working for the Petroleum Research Institute in Venezuela

1994 **Gutiérrez** becomes deputy director of the Research and Special Programs branch of the U.S. Department of Transportation

1947 **The Sol** family moves to Montgomery County, Maryland, where Ana attends school.

1990 Gutiérrez is elected to the Montgomery County School Board.

When Ana Sol Gutiérrez ran for a seat on the Montgomery County, Maryland, school board, it was her first try for public office. During her campaign, Gutiérrez expressed concern and support for minorities and bilingual education. She was elected to the board in November, 1990. Gutiérrez became the first Latin American member of a Maryland school board. She was also the first Salvadorean elected to any public office in the United States.

Gutiérrez was born Ana Emma Sol in El Salvador in 1942. Her mother, Ana Peréz, was a homemaker. Her father, Jorge Sol-Castellanos, was a diplomat. He was El Salvador's first finance minister and was the director of the World Bank and the International Monetary Fund.

Gutiérrez moved to Montgomery County, Maryland at age 5. There she attended public schools. Eventually, she graduated from Chevy Chase Senior High School. After graduating from high school, Gutiérrez moved to Switzerland, where she studied chemistry and liberal arts at the University of Geneva. She then returned to the

United States, earning a bachelor's degree in science at Pennsylvania State University. She later received a master's degree in management and computer sciences from the American University in Washington, D.C.

After receiving her master's degree, Gutiérrez moved to South America. She worked as a professor of mathematics and computer sciences at a university in Bolivia. Gutiérrez later moved to Venezuela. Here, she taught courses in engineering and computer science

While living and teaching in Venezuela, Gutiérrez also worked for the Venezuelan Petroleum Research Institute. In 1976, she designed a computer system for the company. She also worked as a consultant for South American companies in Bolivia, Venezuela, and Peru.

She later returned to the United States to work as an aeronautical engineer. Until 1992, she was senior systems analyst for a company in Seabrook Maryland. Then she worked for the Federal Aviation Administration (FAA). This agency is responsible for monitors air traffic and safety.

Gutiérrez has also worked as a consultant on programs for NASA and the Goddard Space Center. In 1994, She became deputy director of the Research and Special Programs branch of the U.S. Department of Transportation.

While working as an aerospace engineer, Gutiérrez became interested in educational issues. She views education as the instrument that individuals, as well as the entire Hispanic community, need to succeed. Gutiérrez became an active member of the Montgomery County Parent Teacher's Association (PTA). She was also involved in other organizations within the community.

Gutiérrez's main concerns about education dealt with minority and bilingual education. Bilingual education is a teaching program that uses two languages. One of the languages used is the main language of the society. The second language is the primary or native language of the student. In the Maryland school district served by Gutiérrez, Spanish was the native language of the growing Latino population. Gutiérrez pointed out that because minority enrollment in Montgomery County had increased to more than 36 percent of the county's total 103,000 students in 1990, the time had come to focus on minority issues.

Because of her computer background, Gutiérrez understood the importance of computer literacy. She felt that an increase in school board funding was needed to provide computer-assisted instruction in the schools. Her beliefs helped her win a seat on the Montgomery County School Board.

Gutiérrez is well spoken and highly respected. She is often invited to speak before groups on controversial issues. One such issue is a proposal to make English the official language in various states around the country. Gutiérrez opposes this proposal.

Bilingual education has become controversial. Many people believe bilingual programs are successful such programs should be continued. Others believe bilingual education programs do not meet their goals and that such programs should be stopped. As recently as 1996, some states have begun to reject the idea of bilingual education. In these states, lawmakers have voted to make English the official language of the state. This movement is bringing an end to many bilingual education programs.

Gutiérrez's work in education goes beyond her school board service. She served on the academic review board of Maryland Senator Barbara A. Mikulski. She has also been a member of the Governor's Commission on Hispanic Affairs and was named the National School Board's Hispanic Caucus vice president for the northeast region. She also serves on the National Science Foundation Advisory Committee for the Minority Education Project.

In 1994, Gutiérrez became deputy administrator of the Research and Special Programs Administration (RSPA). The agency is part of the U.S. Department of Transportation. This agency monitors the nation's transportation industry and provides training and assistance in matters of transportation safety. It sets standards, conducts research, and protects the public from dangers involved with transport of hazardous materials.

Vocabulary

A SYSTEMS ANALYST studies the procedures of an activity to define its goals and to make sure the activity is being done efficiently.

HAZARDOUS MATERIALS are substances that pose a threat to human health and the environment because they are poisonous, corrosive, flammable, explosive, or radioactive.

It's YOURTURN

Hands-On Activity

ANALYZING TRENDS IN HOME COMPUTER USE

MATERIALS (per student)

graph paper, pencil

BACKGROUND INFORMATION

Each year, more and more schools make use of computer assisted instruction and offer courses in computer sciences. Such courses include keyboarding, word processing, programming, and computer languages. At the same time, computer use in the home is also becoming more common. Today, home computers are used for education, business, and entertainment.

GROWTH OF HOME COMPUTER USE

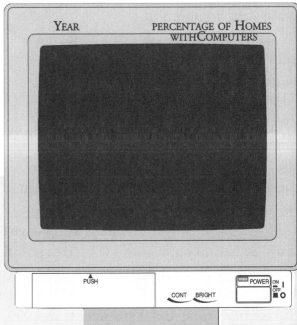

PROCEDURE

1. Examine the table provided.

2. Use the data in the table to prepare a line graph. Plot Number of Home Computers along the Y (vertical) axis. Plot Year along the X (horizontal) axis.

3. Create a title for your completed graph.

ANALYZE AND CONCLUDE

Answer the following questions on the lines provided.

1. What information does your graph show?

   _____

2. In what year does the lowest percentage of home computers appear? The highest percentage?

   _____

3. By what percentage did the number of home computers increase from 1983 to 1987? From 1983 to 1995?

   _____

4. About what percentage of homes had computers in 1990?

   _____

5. About what year did 33 percent of homes have computers?

   _____

6. From the curve on the graph, what can you conclude about the percentage of homes with computers since 1983?

   _____

   _____

Ana Sol Gutiérrez

Think WORK ACT

CRITICAL THINKING *Answer the following questions in complete sentences.*

1. How have Ana Sol Gutiérrez's career interests played a part in her involvement in education?

_____

_____

2. How can computers be used to improve education in the classroom?

_____

_____

3. How might each area of study of Ana Sol Gutiérrez be useful in her work as an aeronautical engineer?

_____

_____

GOING FURTHER
Complete three of the following.

BUILD YOUR PORTFOLIO

Interview a school board member to find out why he or she became involved in community service. Write an article based on your interview. Include a copy of the interview and your article in your portfolio.

USING COMMUNITY RESOURCES

Attend the next meeting of your school board. Take notes about what issues were discussed at the meeting and what views people expressed about these issues. Present your findings in an oral report to your class.

JOURNAL ACTIVITY

In your journal, write about an issue you think needs to be addressed by your school. Explain why you chose this issue and how you would like to see it handled.

COOPERATIVE LEARNING

Do research on and organize a debate about bilingual education and its presence in public schools. One group should present ideas that support the need for bilingual education. Another group should present ideas to oppose this issue. Have a third group monitor the debate and vote on which side presented the best argument.

RESEARCH AND REPORT

Do research to find out what states have passed laws making English their official language. Find out why the law was passed in each state and what effects the change has had.

ROMY LEDESMA

1936 -

1936 *Romy Ledesma* is born in El Paso, Texas.

1971 *Ledesma* receives her B.S. in biology from the University of Texas.

1987 *Ledesma* receives her Ph.D. in college and university administration.

1997 *Ledesma* is working on her second Ph.D. This one is in plant ecology.

1963 *Ledesma* receives her business certification from the International Business College.

1977 *Ledesma* receives her M.S. in biology.

1989 *Dr. Ledesma* establishes an MRCE at the University of Texas in El Paso.

There are more than 3600 colleges and universities in the united states. Of these, only eight have a Materials Research Center of Excellence (MRCE). MRCEs are nationally funded centers. Their goal is to support research efforts and conduct programs that encourage minority and female high school students to enter the fields of sceince and engineering. Of the eight MRCEs, two were founded at the University of Texas at El Paso by Dr. Romy Ledesma, a Mexican American plant ecologist and industrial engineer. Ledesma is the only woman in the United States to administer an MRCE.

Maria Romelia (Romy) Ledesma was born in El Paso, Texas in 1936. She entered elementary school speaking only Spanish, her native language. However, all of her classes were taught in English. Despite this language barrier, Romy was an enthusiastic student who learned quickly.

She soon moved ahead of her English-speaking classmates.

Romy's achievements and academic abilities did not go unnoticed. During her elementary and high school years, she was moved ahead of her grade several times. As a result, she graduated from high school three years ahead of schedule.

After graduation, Ledesma's education seemed to come to an abrupt end. According to Latino tradition at the time, young girls did not leave home to attend college. Thus, Romy remained at home with her family. She was married by age 16. By the time she was 23, she was the mother of four children—two daughters and two sons.

Shortly before her children entered school, Ledesma enrolled in the International Business College in El Paso, Texas. At the school, she studied accounting, stenography, and business machines. She received her certification in business, which prepared her for secretarial work in 1963. She then began work as an accountant and secretary.

PLANT ECOLOGIST

In 1967, Romy entered the University of Texas at El Paso. She studied botany, geology, and Spanish. In 1971, she received her B.S. in biology. She then took a job as a research technician and became herbarium curator at the Texas A&M Research Center in El Paso.

While working at Texas A&M, Ledesma continued her education at the University of Texas. She received her M.S. degree in biology in 1977. Her major was in desert plant ecology. After receiving this degree, Ledesma took a job as a graduate teaching assistant at Texas A&M University in College Station, Texas. She taught courses dealing with the native plants of Texas and plant reproduction.

While teaching and taking courses, Ledesma became interested in business administration. She was most interested in college administration. She had studied in this area as an intern at the University of Texas at El Paso in 1983, where she worked in the president's office. The next year, she studied management at the Office of Budget and Management at the Texas State Governor's Office. In 1987, Ledesma received her Ph.D. in college and university administration from Texas A&M University. While working on this degree, she also minored in industrial engineering.

In 1989, Ledesma established the MRCE at the University of Texas at El Paso and served as its coordinator. As project coordinator, she managed the project's budget and supervised the activities of faculty, research, and educational programs. While working with the MRCE, Dr. Ledesma also began working on a second Ph.D. Her area of study for this Ph.D. was in plant physiology and entomology. Entomology is the study of insects.

In 1993, Dr. Ledesma began a new research center. This center was the Border Biomedical Research Center (BBRC). Like the MRCE, the BBRC was also established at the University of Texas at El Paso. One year later, Dr. Ledesma began work with the EPA Americorps program. The program was started by President Clinton. Dr. Ledesma served as the program's manager and coordinator. In this role, she supervised students and others in a project that involved identifying possible sources of pollution to a local groundwater system.

Today, Dr. Ledesma remains active in several educational and professional causes. She has been well recognized for her work in these areas. In 1992, the National Research Foundation honored Ledesma for her commitment and outstanding work at the MRCE. The same year, for her work as a scientist and role model to minorities and females, she was inducted into the El Paso Women's Hall of Fame. She was also elected to the executive committee of the Texas Association of Chicanos in Higher Education.

In addition to her roles as an educator and role model. Dr. Ledesma has an active family life. She raised her four children to adulthood. In turn, she is the proud grandparent of 15 grandchildren (8 girls and 7 boys) and the great grandparent of three children (2 girls and 1 boy).

VOCABULARY

A HERBARIUM is a collection of dried plants..

ECOLOGY is the branch of science that studies the relationships between living things and their surroundings.

PHYSIOLOGY is the study of how the cells, tissues, and organs of a body function.

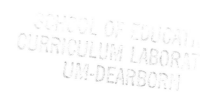

It's YOURTURN

Hands-On Activity

SOCIAL INSECT BEHAVIOR

MATERIALS (per group of 5)
ant-farm kit

BACKGROUND INFORMATION

The activities of insects differ according to the species. Some insects lead solitary lives, associating with other insects only during mating or when obtaining food. Other insects, such as bees, wasps, termites, and ants live in large groups called colonies. Within each colony, each group member has a specific job. The division of labor ensures that the group survives through reproduction, obtains food, and that the colony is defended. Within an ant colony, there is usually one or more females called queens. The fertile males of the colony are called drones. The job of the drones is to mate with the queens. There are also soldier ants that guard the colony. Other ants called workers perform such tasks as getting food and caring for the eggs and young.

PROCEDURE

1. Obtain and set up an ant farm according to the instructions provided in the kit.
2. Observe your ant farm for a few minutes after it is assembled. Sketch the appearance of the farm and identify the different members of the colony on your sketch.
3. Each day for two weeks, repeat step 2. Look for and report on any differences in the sizes, shapes, and body colors of the ants in your colony. Feed the ants lettuce, carrots, potatoes, and bread crumbs or the prepared food source that came with your kit.

ANALYZE AND CONCLUDE

Answer the following questions on the lines provided.

1. What kinds of activities did you observe taking place in the colony over the two-week period?

    _____

    _____

2. What differences are observable in the different members of the colony?

    _____

    _____

3. Which members of the colony were you able to identify? How?

    _____

    _____

    _____

4. Can you compare any of the activities you observed to similar activities in other animals or humans? Explain.

    _____

    _____

    _____

    _____

Romy Ledesma

Think WORK ACT

CRITICAL THINKING *Answer the following questions in complete sentences.*

1. Which of Ledesma's areas of study would have been most helpful in her job as an herbarium curator?

   _____

   _____

2. How might Ledesma's early experience working in business have helped her in her role as an MRCE project administrator?

   _____

   _____

   _____

3. In what ways is the work done by plant ecologists and that of entomologists related?

   _____

   _____

   _____

GOING FURTHER

Complete three of the following.

BUILD YOUR PORTFOLIO

Survey your community to find out what insects or plants live there. Make sketches or take photographs of the insects or plants you observe for inclusion in your portfolio. Use reference materials such as field guides to identify the organisms.

ALTERNATE ASSESSMENT

Conduct research to find out the roles of other colonial insects such as termites and bees. Compare the roles of the individuals in the colonies you investigate to those you studied in your ant colony.

JOURNAL ACTIVITY

Write a haiku or other type of poem about some kind of insect. Try to make your poem as scientifically accurate as possible.

COOPERATIVE LEARNING

Obtain labels from a variety of insecticides or other pest control products. In your group, study the labels to find out what organisms each product is used to control, how the product is applied, what precautions need to be taken when working with the product, and how often the product must be used. Combine your data in a group table.

RESEARCH AND REPORT

Do research to identify five occupations that involve work with either plants or insects. Conduct additional research on one of the occupations you identify to find out what the career involves and what qualifications are needed to prepare for a career in this area. You may wish to conduct an interview with someone who works in their field as part of your research.

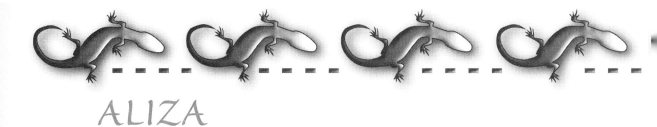

ALIZA

LIFSHITZ

1947 -

1947 *Aliza Lifshitz* was born in Mexico City, Mexico.

1976 *Aliza Lifshitz* receives her M.D. from a college in Mexico and then travels to the U.S. to complete her medical training.

1992 *Dr. Lifshitz* receives a Distinguished Contributor to Social Welfare award from the University of Southern California Los Amigos de la Humanidad School of Social Work.

1969 *Aliza Lifshitz* receives her B.S. degree from a college in mexico.

1991 *Dr. Lifshitz* receives a Physician o the Year award from the Multicultural Area Health Education Center.

Radio and television can reach large audiences. For this reason, these media are often used to broadcast messages and information about public safety and personal health. Many networks now have physicians who regularly appear as hosts or reporters. Latin American physician Dr. Aliza Lifshitz is such a person. She works for a television network in Los Angeles, California that targets Spanish-speaking viewers.

Aliza Lifshitz was born in Mexico City, Mexico. Her father was of Jewish heritage. He made his living in Mexico working as an engineer. He was also a talented classical pianist who trained and received a diploma from Mexico's Music Conservatory. Aliza's mother was Mexican and Catholic. She made her living as an artist.

Aliza attended the American elementary school in Mexico City. There she learned English and met children of diverse cultures. Aliza's later education took place at Jewish and Catholic institutions. She first attended Mexico City's

Jewish High School. She later went to the Catholic Institute of LaSalle.

After high school, Aliza went to the Colegio Israelite de México. She earned her bachelor's degree with honors from the schoo in 1969

Aliza was greatly influenced by h parents' backgrounds in the arts an sciences. She wanted a career that would allow her to be both creative and helpful to the community. She viewed medicine as an art that was based on a science. Thus, she decided to become a physician.

To reach her goal to become a physician, Aliza went to the Universidad Nacional Autonome de México. She graduated from the school with honors in 1976.

After earning her medical degree, Aliza traveled to the United States to complete her medical training and to do her residency. In the United States, she attended Tulane University in New Orleans, Louisiana. She then went to the University of California at San Diego.

After earning her medical degree, Dr. Lifshitz opened a private practice in Los Angeles. She specialized in internal medicine, clinical pharmacology and endocrinology.

Dr. Lifshitz has developed a reputation for being very compassionate. Many of her patients travel long distances to have her treat their illnesses. One reason Dr. Lifshitz has such dedicated patients is because she is willing to take the time needed to be sure her patients understand their health conditions. For example, when treating someone, she describes each step of the treatment process. She also explains why the treatment is needed and what can be learned from ordered laboratory tests. In addition, Dr. Lifshitz does not deny treatment to people who cannot afford medical care. Often, she accepts payment in the form of baked goods, fruits and vegetables, or home made items. Dr. Lifshitz also works with community groups to help people get medical care even if they cannot pay for services.

An area of medicine that is of great interest to Dr. Lifshitz is the treatment of AIDS. Dr. Lifshitz was one of the first Latin American physicians to join in the fight against this disease. She treats many patients who are HIV-positive. She serves on AIDS committees for the Los Angeles Medical Association as well as the California Medical Association. She also works with local, state, and private AIDS groups.

One of Dr. Lifshitz's talents is her ability to reach out to her patients. While appearing on a television show, Dr. Lifshitz received hundreds of calls. She quickly realized how many people TV allowed her to reach. Since then, she has been the health reporter for Univision, a Spanish language television network in Los Angeles. She also produces specials on health topics.

VOCABULARY

AIDS is an acronym for acquired immune deficiency syndrome, a group of illnesses that are caused by a viral attack on the immune system.

HIV is an acronym for human immunodeficiency virus, the virus that causes AIDS.

Dr. Lifshitz often speaks in favor of universal basic health care. She strongly believes that everyone in the United States should be able to receive basic health care no matter what their economic status. She is troubled by the increasing number of working poor in this country who lack medical insurance. She is also concerned about the many retired elderly people who are in need of medical care. Dr. Lifshitz is also an advocate on issues concerning women's health. She believes that as more women enter and succeed in health-care professions, more attention will be paid to women's health-care issues.

Dr. Lifshitz is the editor-in-chief of *Hispanic Physician*. She is also a medical editor of *Más*, a national Spanish-language magazine. Her work with these magazines helps to make sure that Spanish-speaking peoples receive accurate information that can help protect them from infectious diseases.

Dr. Lifshitz is known for her work as a physician and as a reporter. In 1991, she received the Multicultural Area Health Education Center's Physician of the Year Award. The same year, she also received the Commission Femenil's Women Making History Award and its Women in the Health Sciences Award. In 1992, Dr. Lifshitz received the Distinguished Contributor to Social Welfare award from the University of Southern California, Los Angeles-Los Amigos de la Humanidad School of Social Work.

Dr. Lifshitz is also active in many organizations. Among them is the California Hispanic American Medical Association. In 1992, she served as this organization's president.

It's YOURTURN

EVALUATING PUBLIC SERVICE MESSAGES

MATERIALS (per group of 5)

notes from electronic-media survey data table

BACKGROUND INFORMATION

Electronic and print media are powerful means of communication. In many areas, radio, television, and newspapers reach hundreds of thousands or even millions of people daily. In recent years, public service messages, especially those related to AIDS, teen pregnancy, education, drug and alcohol abuse, and smoking have been carried by major television and radio networks. To help make the message more appealing to the intended audience, a well-known radio, TV, or movie personality is often used to deliver the message.

PROCEDURE

1. Create a table with headings as shown below.

2. Each day for one week, survey electronic media listening for public service announcements dealing with health issues. (Each team member should survey a different station or media source.) Record the information you obtain in your table. In the last column, rank how effective you think the announcement was, using a scale of 1-5. Let 5 represent Very Effective and 1 represent Not Effective.

3. After one week, combine the data gathered by all group members in a group table. Use the group table to answer the Analyze and Conclude questions.

ANALYZE AND CONCLUDE

Answer the questions on the lines provided.

1. Which topic was most often presented in the public service messages?

2. At what time of day was the topic from question 1 presented?

3. What percentage of the ads surveyed made use of well-known personalities to deliver a message?

4. Did all of the messages appear to appeal to the same audience? Explain

5. In your opinion, which message was most effective? Least effective? Explain.

6. Which topic not aired should have been given air time? Explain.

Source Surveyed (Radio/TV) Station	Topic	Time of Message	Celebrity used?/ Who?	Message Effectiveness (1-5)

Name _____ Date _____

Think WORK ACT

CRITICAL THINKING *Answer the following questions in complete sentences.*

1. Why do you think television networks often use doctors to give health reports?

_____

_____

2. Of what advantages might speaking more than one language be to a doctor?

_____

_____

3. How might Dr. Lifshitz's practice of accepting gifts for services encourage people to get medical care?

_____

_____

GOING FURTHER

Complete three of the following.

ALTERNATE ASSESSMENT

Survey print media (billboards, magazines, newspapers) looking for public service ads. Record the types of information presented in each ad, the audience you think it was intended to reach, and whether or not a celebrity was used. Prepare a table of the types of information you find similar to the activity on page 26. Rank how effective you think each ad is. Compare the results of the print media survey to those of the radio/T.V. survey.

COOPERATIVE LEARNING

As a group, develop your own public service campaign for a health-care issue that is important to you. Present your group's message in two formats that could be used in print media, radio, or television. Appropriate formats may include posters with drawings or photographs, audiotapes, or videotapes.

RESEARCH AND REPORT

Do research to find out what *cybersurgery* and *virtual reality* are. Prepare an essay to explain how these two ideas can be used to provide medical services.

JOURNAL ACTIVITY

In your journal, list questions you might consider asking if you were going to speak to a doctor on a call-in radio or television program. Explain why each question is important to you.

CONCEPT MAPPING

Identify 10 medical specialties and what is studied in each. Develop a concept map that shows how the specialties are related.

LISSA MARTINEZ

1954 -

1954 *Lissa Ann Martinez* is born in Cherry Point, North Carolina.

1984 *Martinez* becomes the first person to receive a fellowship to study at the National Academy of engineering.

1991 *Martinez* serves on a community task force to change election districts and extend voting rights for local elections to noncitizens.

1976 *Martinez* receives her B.S. in ocean engineering from MIT.

1988 *Martinez* and her husband, Brian Hughes, set up a scholarship fund at MIT to be used by students who want to study engineering.

Water pollution is a worldwide problem. Water pollution occurs when substances that can be harmful to living things are released into bodies of water. Such pollution affects rivers, lakes, ponds, and oceans throughout the world. Water pollution also affects the groundwater systems from which many communities obtain their drinking water. One of the major causes of water pollution is ocean dumping. Ocean engineers try to find ways to prevent pollution of Earth's oceans. Often they work with agencies such as the Environmental Protection Agency (EPA) and the U.S. Coast Guard to deal with ocean dumping. One such engineer is Latin American Lissa Ann Martinez.

Lissa Martinez was born in Cherry Point, North Carolina in 1954. Her parents were Molly and Edmund Martinez. Lissa's father was in the military; he was a member of the Marine Corps. Because his assignments changed often, the family was required to move frequently.

For most children, frequent moving from one place to another on a regular basis is an unhappy experience. Constant moving makes it difficult to make and keep friends and also disrupts schooling. However, Lissa Martinez credits the constant moving of her family with helping in her education. She has stated that because she moved frequently, she never had to deal with teachers who may have had lower expectations of her because she was Hispanic.

When Martinez was in high school, one of her counselors suggested that she apply to the Massachusetts Institute of Technology (MIT) in Boston. Martinez took the advice and received a full scholarship from the school. She began her studies at MIT after graduating from high school.

At MIT, Martinez studied ocean engineering. She received her B.S. in this field in 1976.

However, after graduating from college, Martinez discovered there were few jobs in her

Lissa Martinez

field. She took a job with the Maritime Administration. This agency is involved in shipbuilding and ocean transport. After two years with the Maritime Administration, Martinez returned to MIT to continue her studies. She earned her master's degree in technology and public policy with a major in ocean engineering.

After receiving her M.S. degree, Martinez went to work as an engineer for the United States Coast Guard in Washington, D.C. The main roles of the U.S. Coast Guard are to keep the ocean areas surrounding the U.S. safe and free of pollution. With the U.S. Coast Guard, Martinez worked in the areas of safety and pollution prevention on ocean vessels. U.S. Coast Guard rules require that safety and pollution devices on ships meet certain standards. Martinez and other engineers inspect and test equipment to decide which equipment meets these standards.

In 1984, Martinez received a fellowship to study at the National Academy of Engineering. This academy is afiliated the National Academy of Sciences. Martinez was the first person to be selected for this honor.

While at the academy, Martinez was asked to speak at an engineering conference in Puerto Rico .There, she learned that the school of engineering in San Juan was discouraging students from majoring in electrical engineering. Students who were already in the program were being forced to drop out. The school was facing overcrowding in this major. To solve the problem, they were trying to lower attendance in the program.

The problem facing engineering students in Puerto Rico angered Martinez. She began to speak out and encourage Latin Americans to become

Vocabulary

GROUNDWATER is water that collects below the soil's surface in rock and soil layers. Groundwater is often the source of water for a community.

engineers. She also decided she must do more to help young people who wanted to become engineers reach their goal.

In 1988, Lissa and her husband Brian Hughes set up a scholarship fund at MIT. Hughes works for at a company that designs rocket engines. The couple donated more than $200,000 in stock to the scholarship fund. The fund helps financially needy engineering students at MIT by providing them with grants to help pay for their education costs.

Martinez's role in helping others has not been limited to the creation of the scholarship fund at MIT. She also was the founding member of the Washington, D.C. chapter of the Society of Hispanic Professional Engineers (SHPE). This group works to make others more aware of the important work done by engineers. In addition, the group speaks with young people to encourage them to pursue careers in engineering.

In 1991, Lissa Martinez also became active in her community. She served on a 15-member task force in her hometown of Takoma Park in Maryland. The task force was set up to create six new voting districts in the community. In addition, the task force was able to have voting rights in city elections extended to all members of the community, regardless of whether or not they were citizens of the United States.

Today, Martinez lives in Maryland with her husband Brian Hughes and their two sons. She has opened her own business outside Washington, D.C. The business is a consulting firm that advises the EPA and the U.S. Coast Guard about ways to control and reduce ocean pollution.

It's YOURTURN

LEACHING AND GROUNDWATER

MATERIALS (per group of three)

3 styrofoam cups, paper clip, sand, gravel, soil, potassium permanganate crystals, spatula, water

SAFETY

Wear safety goggles and a laboratory coat or apron throughout this activity. Use a spatula to distribute the potassium permanganate crystals. Use care when working with the paper clip to avoid poking yourself.

BACKGROUND INFORMATION

In many areas, drinking water is obtained from natural springs or wells of the groundwater system. The groundwater system gets its water supply from rainwater that seeps through soil and rock layers. As rainwater moves down through soil layers, it is filtered naturally before it enters the groundwater system. At the same time, this water picks up dissolved minerals in soil in a process called leaching. The leaching of agricultural pollutants (pesticides and fertilizers) and toxic wastes (motor oil, acids, paints) that have been dumped or spilled over large land areas threatens the quality of some well water. This leaching process can be modeled using water to represent rain and a coloring agent to represent a layer of pollutants buried beneath the soil's surface.

PROCEDURE

1. Open the paper clip to form a wire. Use the wire to poke about 5 holes in the bottom of one cup.
2. Line the bottom of the cup with the holes with about 2.5 cm (1 in.) of coarse gravel.
3. Use the spatula to sprinkle a thin layer of potassium permanganate crystals over the gravel layer.
4. Cover the crystals with about 2.5 cm (1 inch) of sand. Cover the sand with about 2.5 cm (1 inch) of soil.
5. Place the cup with the soil on top of the open end of an empty cup.
6. Half-fill the remaining cup with water. Slowly pour the water from this cup into the cup containing the soil.
7. Separate the cup containing the soil from the cup beneath it. Observe the water that has collected in the bottom cup.

ANALYZE AND CONCLUDE

Answer the following questions on the lines provided.

1. In your model, what does the water represent?
   _____

2. What does the soil, sand, and gravel represent?
   _____

3. What do the crystals of potassium permanganate represent
   _____

4. What happened to the water as it passed through the materials in the cup? What does this process represent?
   _____
   _____

5. What is one way that groundwater can become polluted?
   _____
   _____

Lissa Martinez

Think WORK ACT

CRITICAL THINKING *Answer the following questions in complete sentences.*

1. What do you think ocean dumping is? What might be some sources of ocean pollutants?

_____

_____

_____

2. Why is it important that ocean dumping be controlled?

_____

_____

_____

3. How might improperly operating equipment on ships contribute to ocean pollution?

_____

_____

GOING FURTHER

Complete three of the following.

BUILD YOUR PORTFOLIO

Survey your community looking for areas that show evidence of water pollution or areas that may be at risk of pollution. Make a videotape or photo essay of your findings.

ALTERNATE ASSESSMENT

Using cooking oil, water, and other materials you choose, demonstrate several ways one might remove or separate spilled oil from water.

USING COMMUNITY RESOURCES

Write a letter or meet with a member of the U.S. Coast Guard to find out what role this agency plays in protecting the ocean environment from pollution.

RESEARCH AND REPORT

Conduct research to find out how oil spills and other forms of water pollution affect birds, aquatic mammals, and fishes.

COOPERATIVE LEARNING

Research five major oil spills. Identify where and when each spill occurred, how much oil was spilled, and the effect it had on wildlife and the environment. Combine your individual data with that of other group members in an illustrated report.

ANTONIA MAURY

1866 - 1952

1866 Antonia Maury is born in Cold Spring, New York.

1888 Maury begins work at the Harvard Observatory.

1896 Maury leaves the Harvard Observatory. She spends the next 22 years working as a teacher and lecturer.

1935 Maury retires from the Harvard Observatory and becomes curator of the Draper Park Observatory Museum.

1887 Maury graduates from Vassar College.

1889 Maury observes that the star known as Mizar produces two different spectra, an observation supporting the hypothesis that Mizar is part of a binary star system.

1918 Maury returns to the Harvard Observatory.

1952 Maury dies at the age of 85 in Dobbs Ferry, New York.

Can you find the Big Dipper in the night sky? The star in the center of the Big Dipper's handle is named Mizar. However, Mizar is not really a single star. It is actually two stars located so near each other they appear to be only one star. Mizar was identified as part of a binary, or double star, group in 1650. However, it was not until 1889 that the existence of binary stars was proved. This proof was provided by astronomers Edward C. Pickering and Antonia Maury.

Antonia Maury was born in Cold Spring, New York in 1866. She was the oldest of three children born to Reverend Mytton Maury and his wife Virginia Draper Maury, a woman of Portuguese descent.

Maury's family was greatly involved in the sciences. Antonia's maternal grandfather, John William Draper, was a physician and amateur astronomer. He became the first person to make a photograph of the moon. The photograph was called a daguerreotype. Draper's son Henry (Antonia's uncle) was a physician and amateur astronomer. Henry Draper was the first person to photograph a star's spectrum to reveal the lines it

produced. Much of the work later done by Antonia Maury made use of the method of gathering spectra that was developed by her uncle and other astronomers who built upon his work.

Interest in the sciences was not limited to the mother's side of the family. In addition to being a minister, Antonia's father was a naturalist and an editor of a geographic magazine. The family interest in science was passed on to the Maury children. Antonia became a noted astronomer; her sister Carlotta became a noted paleontologist.

Antonia Maury received her early education from her father. She went on to Vassar College in Poughkeepsie, New York. At Vassar, Antonia studied astronomy under Maria Mitchell. Mitchell is generally recognized as the first female astronomer in the United States.

In 1887, Maury graduated from Vassar with honors in astronomy, physics, and philosophy. One year later, Maury went to work at the Harvard Observatory. Part of the observatory, the Henry Draper Memorial, was provided in honor of Maury's uncle in 1886.

Antonia Maury

At the observatory, Maury worked for Edward C. Pickering. He had developed a system for classifying stars by their spectra. Spectra are the bands of colors that form when light passes through a prism. Each star produces a unique spectrum. Thus, a spectrum can be used to identify a star. Maury's job was to use Pickering's system to classify stars. Maury obtained the spectra by photographing starlight as it passed through a prism placed in front of her telescope.

From her observations of more than 4,000 spectral photographs, Maury improved Pickering's classification system. Her studies also produced much data. In 1889, she observed that Mizar formed two different spectra. These data supported a 1650 hypothesis that Mizar was part of a binary star group. In addition, these data were used and built upon by other astronomers.

Maury paid great attention to her work. As a result, her research was very time-consuming. Pickering believed Maury needed to much time to complete her work. Maury and Pickering also had many personal conflicts. Some of the conflicts arose from the changes Maury made in Pickering's classification system. Other conflicts arose because Maury's family, who had strong ties to the observatory, became involved in the problems between Maury and Pickering.

As a result of their constant conflicts, Maury began to take a lot of time off from work. Finally, in 1896, she left the observatory. For the next 20 years, she worked as a teacher and lecturer at Cornell University and several other colleges.

Pickering retired from the Harvard Observatory in 1918. Following his resignation, Maury returned to the observatory. She remained at the observatory, continuing her work on spectral analysis of binary stars, until 1935. During this time, Maury also determined the orbits of several binary stars.

After leaving the observatory, Maury served as curator of the Draper Park Observatory Museum. She held this post for three years. During this time, and until a few years before her death, Maury visited the Harvard Observatory each year to check the accuracy of her previous work.

Astronomy was not Maury's only area of interest in the sciences. Maury was very involved in ornithology. She was also a naturalist and an active conservationist. One of her conservation projects involved helping to preserve the giant redwood forests in the western United States.

Maury was a member of the American Astronomical Society, the Royal Astronomical Society, and the National Audubon Society. In 1943, she was awarded the Annie J. Cannon Prize from the American Astronomical Society. Cannon was a woman astronomer who became known for her work in star classification using spectral analysis. Ironically, the Cannon Prize was awarded to Maury in recognition of her revision of Pickering's classification system—one cause of the conflicts between the two scientists.

After Maury gave up her post as curator of the Draper Park Observatory Museum, she moved to the former home of the Drapers in Hastings-on-Hudson. After becoming ill, Maury was moved to a hospital in Dobbs Ferry, New York. Maury died at the hospital on January 8, 1952.

> **VOCABULARY**
>
> A BINARY STAR is a star system made up of two stars that revolve around each other.
>
> A PALEONTOLOGIST is a scientist who studies ancient life forms from their fossils.
>
> ORNITHOLOGY is the branch of biology dealing with the study of birds.

It's YOURTURN

SPECTRAL ANALYSIS OF LIGHT

MATERIALS (per student)

metallic jar lid with about a 4" diameter, sheet of unlined white paper, 2 small candles, safety matches, diffraction grating/spectroscope, colored pencils

MATERIALS

Use caution when lighting the candles with the matches. Tie back loose hair and secure loose clothing. Keep hair and clothing away from the open flame of the candles.

BACKGROUND INFORMATION

Almost half of all stars seen from Earth exist as pairs rather than as single stars. These stars are known as binary or double stars. Some binary stars are positioned so closely together that they cannot be seen as double stars. By allowing the light from these stars to pass through a prism, spectra are produced. Spectra produced by close binary stars can be used to identify each star.

PROCEDURE

1. Place the lid on your desk screw side up. Secure 2 candles onto the lid as demonstrated by your teacher.

2. Rotate the lid until the candles appear to be lined up one behind the other. (When lined up properly, one candle will be hidden behind the other.)

3. Light one candle and observe its flame through the diffraction grating. Using a sheet of white paper and colored pencils, sketch what you see.

4. Light the second candle. Rotate the lid slightly so that you can just see or make out the flame of the second candle.

5. Observe both flames through the diffraction grating. Sketch your observations on the sheet of paper.

6. Observe the candles again through the diffraction grating. Move the diffraction grating about 1 foot from the candles and observe the candles again. Make a second observation moving the diffraction grating approximately 1 yard from the candles. Note how the observations differ.

ANALYZE AND CONCLUDE

Answer the following questions on the lines provided.

1. What colors did you observe in each candle's spectrum? How did the zones of color compare in size?

   _____

   _____

2. How many spectra did you observe when you looked at the flame of only one candle? When you observed both candles?

   _____

3. How did the spectra formed by both candles differ from that of the single candle?

   _____

4. How did distance affect the spectra of the candles? How do you think distance might affec the spectra of binary stars?

   _____

   _____

Antonia Maury

Think WORK ACT

CRITICAL THINKING *Answer the following questions in complete sentences.*

1. How can spectral analysis be used to determine the distance between the stars in a binary star group?

_____

_____

_____

2. How is the spectrum of a star similar to a fingerprint?

_____

_____

3. What role might Antonia Maury's family have played in her career choice?

_____

_____

_____

GOING FURTHER *Complete three of the following.*

BUILD YOUR PORTFOLIO

Antonia Maury spent much of her leisure time studying birds. Survey your community to find out what kinds of birds live there. Sketch, photograph, or videotape the birds you observe. Place your completed project in your portfolio.

PERFECT YOUR SKILL

Conduct library research to find out more about daguerreotypes and other methods of photography. Create a time line to show how the ways in which photographs are produced have changed over time.

JOURNAL WRITING

In your journal, write a poem, rap song, or some other creative lines that describe a star and its light. Try to make your verse as scientifically accurate as possible.

COOPERATIVE LEARNING

As a group, research the various constellations and star groups present in the night sky during a particular season. Create a display that identifies at least five star formations. Include captions for the formations you chose explaining what the formation represents and how it got its name.

RESEARCH AND REPORT

Do library research to find out about the work of Annie Jump Cannon. Find out how the work done by Cannon was similar to that done by Antonia Maury and why Cannon was sometimes called the "census taker of the sky." Write a report of your findings.

YNES

MEXIA

1870 - 1938

1870 *Ynes Mexia* is born in Washington, D.C.

1907 *Mexia* remarries, but the marriage soon ends in divorce.

1925 At age 55, *Mexia* makes her first major plant-collecting excursion to western Mexico. She makes many more excursions over the next 12 years.

1938 *Mexia* die of lung cancer in San Francisco hospital.

1897 *Mexia,* while living in Mexico marries Herman Lane, who dies seven years later.

1921 *Mexia,* now living in California, enters the University of California at Berkeley as a special student.

1937 At age 67, *Mexia* continues her education at the University of Michigan.

In many areas of the world, tropical rain forests are being destroyed.as land is cleared to make room for housing, roads, or for farming. Environmentalists fear this destruction will bring about the extinction of many plant and animal species. Many rain forest plants contain chemicals that are used to treat illnesses. Medicines now made from these plants include taxol, a cancer fighting drug, and quinine, which is used to treat malaria. Many other medicines may be hidden in these plants, but if rain forests are destroyed, the medicines in these plants will go undiscovered.

The uses of plants to make medicines is a topic of much research. However, such research is not new. Ynes Mexia, a Mexican American naturalist and explorer, was an early pioneer in this field.

Ynes Mexia was born in the Georgetown section of Washington, D.C. in 1870. Her father, Enrique Mexia, worked for the Mexican government. Her mother, Sarah Wilmer, raised Ynes and six children from a previous marriage.

Ynes had a troubled childhood that was marke by frequent moves. When Ynes was 3, her paren marriage broke apart. Ynes moved w her mother and siblings to LimestoneCounty, Texas Her family lived on land that was owned by the Mexia family. Ynes remained in Texa until she was 15.

In 1886, Sarah and her children moved to Philadelphia and then to Ontario, Canada. During this time Y attended private Quaker schools. The family moved yet again that same ye This time they went to Maryland, th home state of Ynes' mother.

In Maryland, Ynes attended St. Joseph's Academy in Emmitsburg ur 1887. She then went to Mexico City, where she lived for 10 years in her father's home.

In 1897, at age 27, Ynes married Herman Lane— Spanish-German merchant. The marriage ended when Ynes' young husband died in 1904. Three years later, Ynes married Augustin A. de Reygados. Reygados had worked for Ynes. This marriage quickly ended in divorce.

Mexia moved to San Francisco, California, in 1908. Her divorce had left her depressed and withdrawn. Finally, in 1920, Ynes Mexia developed a new interest—flowers. Mexia began to explore this interest through field trips with the local Sierra Club.

In 1921, at age 51, Mexia was admitted to the University of California at Berkeley. She became interested in the natural sciences. After completing a course in flowering plants, Mexia decided botany would shape her future. At a time when most people begin to think about retirement, Ynes Mexia's life as a naturalist and explorer was about to begin.

Mexia entered her new life with enthusiasm. At age 55, having completed three years of college, Mexia made her first major excursion. In 1925, she traveled through western Mexico. In spite of a fall that left her with fractured ribs and an injured hand, she collected hundreds of plant specimens. One, *Mimosa mexiae,* was named for her by botanist Joseph Nelson Rose.

Mexia returned to western Mexico again in 1926. During this seven-month trip, she collected and preserved more than 33,000 plant specimens. Mexia received her training in the art of collecting from her friend and mentor Alice Eastwood, the noted botanist.

Mexia continued her study of plants by exploring the slopes of Mt. McKinley in Alaska. This was followed by a trip to Brazil and a 30-month excursion into the rain forests of Peru and the Amazon. Mexia described her Amazon adventure in, "Three Thousand Miles Up the Amazon." The article appeared in the *Sierra Club Bulletin*.

During her trip, Mexia traveled by ship, dugout canoe, and balsa raft. She lived on beaches under mosquito netting and bartered with local peoples for food. On the trip, Mexia collected and classified plants, insects, and birds. She collected more than 65,000 plant specimens.

At 62 , Mexia remained passionate about her work. In 1933, she traveled with Alice Eastwood to California, Nevada, Utah, and Arizona. On the trip, the women collected, described, and photographed plants. More than 100 species were collected; one was new to science.

From the American West. Mexia journeyed into the rain forests of Peru, Ecuador, Chile, and Argentina. In Ecuador and Columbia, she looked for Cinchona, the tree whose bark is the source of quinine. She was also successful in her search for the plant sources used by natives as fish poisons.

In 1937, Mexia again returned to the United States.and at 67, began to take outdoor classes at the University of Michigan. The following year, she set out again on a collecting expedition to southwestern Mexico. Several months later, Mexia began to suffer from severe stomach pains. The pain forced her to return to San Francisco. She carried with her more than 13,000 plant specimens. Once back in the States, Mexia was immediately hospitalized. She died of lung cancer on July 12, 1938.

During the course of her work, Mexia collected more than 150,000 plant specimens. She discovered one new genus and more than 500 new plant species. Some of these plants were named in her honor. In addition to describing and classifying new plants, Mexia made many drawings that showed the plants in their natural habitats. Her accurate observations and descriptions confirmed and built upon the work of earlier botanists and naturalists. Her field notes are kept at the University of California at Berkeley.

Vocabulary

A GENUS is a group of several species that share many characteristics.

A BOTANIST is a scientist who studies plants.

It's YOURTURN

IDENTIFYING TREES

MATERIALS

crayon, colored pencils, 6 sheets of typing paper, metric ruler, notebook or clipboard, field guides to trees

SAFETY

Wear appropriate clothing when working outdoors. Avoid contact with poisonous plants such as poison ivy and sumac.

BACKGROUND INFORMATION

The characteristics of a tree are used to identify the tree. Such characteristics include leaf size and shape, the general shape or silhouette of the tree, whether the tree produces seeds from flowers or cones, and bark color and texture. For example, a maple leaf, which appears on the flag of Canada, has a very familiar shape. Trees that may be identified by their silhouettes or general shape include those in the pine family and the elm. Paper birch, yellow birch, and the sycamore are examples of trees that have a unique color and pattern to their bark.

PROCEDURE

1. Survey your community to observe the different types of trees present.
2. Label each sheet of typing paper as Tree Number 1, Tree Number 2, and so on. Use a different sheet of paper for each type of tree you observe.
3. Locate six different types of trees. For each tree, do the following:
   ▲ Sketch the silhouette or the general shape of the tree.
   ▲ Sketch an outline of a leaf to show its general shape as well as its vein pattern.
   Take measurements of a few leaves, without removing them. Note the sizes (length and width) beside your sketch.
   ▲ Make a tracing of the bark by holding the paper against the tree trunk and rubbing the crayon along the paper. Beside your rubbing, identify the color of the bark and any traits that may be helpful in identifying the tree.
   ▲ Observe if the tree has any flowers, fruits, or cones. Sketch any of these features you observe.
4. Use a field guide and your data to try to identify the trees you observed. Note how well the characteristics you observed compare with those given in the guide.

ANALYZE AND CONCLUDE

Answer the following questions on the lines provided.

1. How many of your trees were you able to identify?

   _____

2. What characteristics were most helpful in identifying each tree?

   _____
   _____
   _____
   _____
   _____

3. Based on information in the field guide, what other data could you have collected that would have been helpful in identifying the trees you observed?

   _____
   _____
   _____
   _____

Think WORK ACT

CRITICAL THINKING *Answer the following questions in complete sentences.*

1. Why do you think Ynes Mexia is considered a naturalist rather than a botanist?

_____

_____

2. The work done by Ynes Mexia was done as field studies. What advantages do you think working in the field has over working in a laboratory?

_____

_____

_____

3. How might the work done by Ynes Mexia be helpful to scientists studying tropical rain forests and their plants today?

_____

_____

GOING FURTHER *Complete three of the following.*

BUILD YOUR PORTFOLIO

Create a pictorial or photo essay of plants common to your area. Include 10 trees, 10 grasses, and 10 flowers. Use library resources to label each plant with its common and scientific names.

JOURNAL WRITING

In your journal, describe which area visited by Ynes Mexia you would most like to explore. Give reasons for your choices. What you would most like to learn about the area?

COOPERATIVE LEARNING

Work with three classmates to create a field guide to the plants in your area. Use library resources to find out about each plant. Include drawings or photographs in your guide. Also include information such as each plant's scientific name, its features, and its uses.

RESEARCH AND REPORT

Use library resources to find out about the drug taxol. Prepare a report explaining how taxol is used and where it comes from.

ANTONIA

NOVELLO 1944 -

1944
Antonia Coello is born in Puerto Rico.

1970 *Coello* receives her M.D. from the University of Puerto Rico at Rio Piedras and marries Joseph Novello.

1990 *Novello* becomes the first woman and first Latino to hold the post of surgeon general.

1965 *Coello* receives her B.S. degree from the University of Puerto Rico at Rio Piedras.

1978 *Novello* begins to work for the NIH.

1993 *Novello* resigns the post of surgeon general. She begins work with the University of Michigan, Georgetow University, and UNICEF.

The position of surgeon general of the United States is one of honor and authority. The role of the surgeon general is to improve public awareness of health problems such as those related to smoking, alcohol, and sexually transmitted diseases. The surgeon general also oversees the President's Council on Physical Fitness and Sports.

In 1989, Dr. C. Everett Koop retired his post as surgeon general. On October 17,1989, President George Bush nominated Dr. Antonia Coello Novello to that position. She was sworn in as the fourteenth Surgeon General of the U.S. in March of 1990. With her appointment, Dr. Novello became the first woman and first Latino, to serve as surgeon general.

Antonia Novello was born as Antonia Coello in Fajardo, Puerto Rico in 1944. She was the oldest of three children of Antonio Coello and Ana Delia Coello. When Antonia was 8 years old, her father died. Thus, she was raised primarily by her mother, who worked as a teacher and later as a principal.

When Antonia was born, she had an enlarge colon. As a result of this medical problem, she spent several weeks each year in the hospital. Despite these hospital stays, Antonia was a good student, who graduated from hig school at 15.

Antonia's mother valued education. She stressed this value to her children. Encouraged by her mother, Antonia attended th University of Puerto Rico. While in college, Antonia continued to suffer from the medical condition she had had since birth. She underwent surgery to correct the problem when she was 18, but th surgery was not a complete success. For the next two years, Antonia suffered complications related to her surgery. The problems were finally corrected afte she had a second surgery at age 20.

Antonia's medical problems interrupted her college education. However, following her secor surgery she returned to college to complete her education. She obtained her B.S. degree from th University of Puerto Rico at Rio Piedras in 1965.

Antonia Novello

After receiving her B.S. degree, Antonia went to the university's medical school. She graduated with her M.D. degree in 1970. The same year, she married Joseph R. Novello, a navy flight surgeon.

After the couple married, they moved to Michigan. Antonia began her internship at the University of Michigan Medical Center. She specialized in pediatrics. In 1971, Novello was named intern of the year by the medical center for her outstanding work. Dr. Novello was the first woman to receive this award.

Dr. Novello began her residency in Michigan. During this residency, Novello and her husband moved to Washington, D.C. Dr. Novello completed her residency at the Georgetown University Hospital.

The Novello's moved to Springfield, Virginia, in 1976. Here, Dr. Novello opened a private practice. However, she gave up her practice two years later to work as a project leader at the National Institutes of Health (NIH).

Novello earned a master's degree in public health at the Johns Hopkins School of Public Health in 1982. From 1982 to 1983, she worked both at the NIH and as a congressional fellow. In this role, she worked with the Senate Committee on Labor and Human Resources. While working with the committee, Novello helped write the National Organ Transplant Act of 1984. The act helped create a network for obtaining organs needed for transplants. She also helped write warning labels for cigarette packaging.

In 1986, Dr. Novello was named deputy director of the National Institutes of Child Health and Human Development. In this post, she focused much of her attention on children with AIDS. Dr. Novello's work in this area got the attention of the White House, resulting in her nomination for the surgeon general's post.

While serving as surgeon general, Dr. Novello made health care issues of children, women, and minorities her top priorities. In 1991, she began a campaign to stop young people from using alcohol. She attacked companies that sell alcohol advertisements that appealed to young people. In 1992, Dr. Novello waged a similar attack on tobacco companies. This attack made headlines again (in 1997) when the attorneys general of several states met with tobacco company executives and got them to agree to stop using such advertising.

During her term as surgeon general, Dr. Novello spoke out against domestic violence. She also spoke against alcohol use by pregnant women. Today, alcoholic beverages carry warnings to pregnant women. The warnings identify conditions that may result if the developing fetus is exposed to alcohol. In addition, she emphasized increasing awareness about AIDS through education.

Like cabinet positions, the appointment of a surgeon general is decided by a sitting president. Thus, in July 1993, President Bill Clinton nominated Joycelyn Elders to the position of U.S. Surgeon General. Dr. Elders became the second woman and the first African American to serve as surgeon general. Since leaving the post of surgeon general, Dr. Novello has worked as a professor at the University of Michigan and the Georgetown University School of Medicine. Dr. Novello has also held a post with UNICEF. In this role, she continues her battles against smoking and substance abuse in young people.

Vocabulary

The COLON is also known as the large intestine. It is part of the digestive system and is responsible for absorbing fluids before the remaining solid products from food are eliminated from the body as waste.

It's YOUR TURN

Hands-On Activity

ANALYZING THE EFFECTS OF ALCOHOL IN THE BLOOD

BACKGROUND INFORMATION

During her term as surgeon general, Antonia Novello focused attention on alcohol abuse among the nation's youth. Because alcohol is legal for adults, many people do not recognize its dangers. However alcohol is one of the most widely used and abused drugs. Pregnant women who drink risk giving birth to babies with fetal alcohol syndrome. Babies born with this syndrome may have physical and mental disabilities. Accidents involving drunk drivers make alcohol the number one cause of death among teenagers. The amount of alcohol in the bloodstream is called the Blood Alcohol Concentration or BAC. A person's BAC can be determined by analyzing air exhaled into a specially designed metering device.

PROCEDURE

1. Carefully study the table. Use the information in the table to complete the Analyze and Conclude.

Effects of BAC Concentration	
BAC (percent)	Effects
0.02 - 0.03	Feeling of warmth and relaxation
0.05 - 0.06	Slight loss of reaction time and coordination
0.08 - 0.09	Loss of coordination, judgement, balance, and speech
0.10 - 0.12	Reaction time slowed, speech slurred, poor coordination
0.13 - 0.15	Seriously impaired vision, speech, balance, and coordination
0.20 - 0.30	Confusion, little control of mind or body
0.40	Unconsciousness
0.50 - 0.60	Coma, respiratory failure, death

ANALYZE AND CONCLUDE

Answer the following questions on the lines provided.

1. What effects will a BAC of 0.02 - 0.03 have on the body?

   _____

2. At which BAC level is a person likely to lose consciousness?

   _____

3. What can you conclude about the effects of alcohol on the body as the BAC increases?

   _____

4. At what BAC can coma, respiratory failure, and death occur?

   _____

5. In many states, a person who drives with a BAC above 0.08 can be arrested and prosecuted. Why is driving with a BAC of 0.08 or higher dangerous?

   _____

   _____

   _____

Think WORK ACT

Name _____   Date _____

CRITICAL THINKING *Answer the following questions in complete sentences.*

1. How is the job of the surgeon general important to society?

   _____

   _____

2. Many people have suggested that the job of the surgeon general be eliminated in an effort to make government smaller and save tax dollars. Do you think the job of the surgeon general should be eliminated for these reasons. Why or why not?

   _____

   _____

3. What qualifications do you think a candidate for surgeon general should have? Explain.

   _____

   _____

GOING FURTHER   *Complete three of the following.*

BUILD YOUR PORTFOLIO

Find the names of five people who have served as surgeon general. Make a time line that shows when each person served in the position.

CONCEPT MAPPING

Find out what processes are involved between the time a person is nominated to the position of surgeon general and the point at which the person takes office. Organize your findings in a concept map.

JOURNAL WRITING

In your journal, describe a health-care issue that was focused on by Antonia Novello that is also important to you. Explain why you selected this issue.

COOPERATIVE LEARNING

As a group, select a public-health issue you think people should know more about. Design a campaign to make people more aware of the issue. Prepare the materials needed to carry out your campaign.

RESEARCH AND REPORT

Choose one surgeon general on which to do more research. Find out who nominated this person to office, what health issues the surgeon general was most concerned with, and how long the person remained in the post of surgeon general. Prepare a report of your findings.

ADRIANA OCAMPO

1955 -

1955 *Adriana Ocampo* is born in Barranquila, Columbia, but later moves to and is raised in Argentina.

1983 *Ocampo* receives her B.S. in planetary geology from the California State University at Los Angeles and begins to work full-time for the Jet Propulsion Laboratory of NASA.

1990 *Ocampo* organizes the Space Conference of the Americas, an event that brings together scientists from around the world.

1970 The *Ocampo* family moves to Pasadena, California.

1989 The *Galileo* spacecraft is launched. *Ocampo* is the science coordinator on the project.

1992 The *Mars Observer* spacecraft, an exploratory mission on which *Ocampo* works, is launched.

About 65 million years ago, dinosaurs vanished from Earth. Scientists have spent years trying to explain this event. In 1975, Harold C. Urey, a Nobel prize-winning geochemist, suggested that a comet may have collided with Earth, causing a dust cloud that blocked out sunlight. More recently, Adriana C. Ocampo, has suggested that a meteorite which struck the Yucatan Peninsula in Mexico may have helped bring about the extinction of the dinosaurs. Adriana Ocampo is a Latin American planetary geologist.

Adriana C. Ocampo was born in Barranquilla, Columbia, in 1955. When she was only a few months old, the Ocampo family moved to Buenos Aires, Argentina. Adriana grew up and attended elementary school in Argentina.

As she prepared to enter secondary school, Adriana was given a series of aptitude tests. The results suggested Adriana would do well in business and accounting. Adriana, however, was more interested in the sciences.

When Adriana was 15, her family moved again, this time to the United States. They settled in Pasadena, California. Here, Adriana attended a local public school. While in high school, Adriana wanted to take a course in auto mechanics. The school would not allow her to do so. Her counselors suggested she take business courses instead. Adriana began to believe that the schools in the United States were no better than those in Argentina in encouraging girls to become whatever they wanted.

Adriana's earlier interest in science and math continued. She especially enjoyed physics and calculus. In her third year of high school, she took part in a school program at NASA's Jet Propulsion Laboratory (JPL) in Pasadena. The program captured her interest. At the end of the school year, Adriana remained with the JPL in a part-time job.

Adriana continued to work at the JPL when she began studies in aerospace engineering at Pasadena City College. However, while in this program, Adriana's interests changed. She decided to study geology. She did so at the California State University at Los Angeles. She received her B.S. degree in geology from the school in 1983.

After obtaining her degree, Adriana turned her 10 years of part-time work at the JPL into a full-time position. She got a job as a planetary geologist. In this work, she has been involved with

Adriana Ocampo

PLANETARY GEOLOGIST

Project Galileo, a probing mission to Jupiter. She has also worked on the Mars *Observer* mission and the *Hermes* mission to explore Mercury.

Ocampo was the science coordinator on *Project Galileo*, which was launched in 1989. The spacecraft reached the planet Jupiter in 1995. *Project Galileo* was the first NASA mission to study Jupiter and its satellites. On this mission, Ocampo controlled a sensing device on the spacecraft. The device, called a spectrometer, identifies and analyzes the gases near a planet's surface. These data gave scientists on Earth information about Jupiter's atmosphere.

The Mars *Observer* spacecraft was launched on September 25, 1992, to explore Mars. It was the first mission sent to Mars since the *Viking* spacecraft of 1975. In August of 1993, NASA lost radio contact with the Mars *Observer*. This event brought the mission to a halt before the probe reached its destination. However, during the Mars *Observer* mission, Ocampo was able to gather some data about the atmosphere of Mars. She also produced a photo atlas of Mar's moon. The data in the atlas was later used on a Russian space mission. To date, the photo atlas is the only work of its kind.

Ocampo has twice applied to NASA for admission into its astronaut program. Both times, her applications were rejected. Still, she is determined to join the ranks of those who have gone into space.

Ocampo believes space exploration will unlock the mystery of Earth's evolution as well as that of its life forms. Her membership in several professional organizations is deeply rooted in learning more about Earth's history and the events that have shaped Earth over geologic time. One such organization, called the Chixulub Consortium, continues to study a crater site in the Yucatan Peninsula. Ocampo and other scientists believe the crater, called , Chixulub, may have

been created when a meteorite struck the area. The impact caused huge clouds of dust and gases. These, in turn, may have produced acid rain, destroying the food supply of the dinosaurs.

When Ocampo was growing up, there were no female astronaut role models to whom she could turn. She credits most of her success to her parents. Throughout her youth, they supported and encouraged her, telling her she could do and be anything. At the same time, they reminded her of the importance of a good education.

As a member of the Society of Hispanic Professional Engineers (SHPE), Ocampo heads the society's international affairs committee. She is also a member of the space committee. In 1987, Ocampo put together a course in Planetary Sciences that was taught in Mexico City. The course was so successful that the United Nations, the Planetary Society (founded in the 1970s by noted astronomer Carl Sagan), and other agencies provided funding for similar programs in Costa Rica, Columbia, Nigeria, and Egypt.

In 1990, Ocampo gathered a group of scientists from around the world to have them share information on space exploration. The event, called the Space Conference of the Americas, met in Costa Rica. Ocampo is also involved with the United Nations and the Planetary Society. She also gives workshops and lectures in space science in Columbia, Costa Rica, and Mexico.

Ocampo belongs to several organizations related to her field of study. She is a member of the Association of Women in Geoscience, the American Institute for Aeronautics and Astronautics, and other professional groups. She is also a spokesperson for the JPL on its mission to create interest in space science and engineering. Ocampo has been recognized for her work in her field and her encouragement of women in the sciences. In 1992, she received the Women of the Year Award in Science by the Commission Femenil.

> ## Vocabulary
>
> A GEOCHEMIST is a scientist who studies the chemical makeup of Earth or other bodies in space
>
> A PLANETARY GEOLOGIST is a scientist who studies the physical features of planets

Name _____   Date _____

It's YOURTURN

IDENTIFYING ELEMENTS USING A FLAME TEST

BACKGROUND INFORMATION

Testing for the presence of various elements can be done in many ways. Optical instruments, such as the spectroscope, are used to analyze a color spectrum that indicates the presence and identity of the elements in stars and the atmospheres of planets. In a similar way, a flame test can be used to determine the presence of metallic ions in a substance. Many metallic ions give off a characteristic color when heated in a flame. The color identifies the presence of these ions.

PROCEDURE

1. Ions of different elements give off light of different colors when burned. These colors can be observed by conducting a flame test. Table 1 identifies the color of the flames given off by different elements when they are burned. Review this information.

ELEMENT	FLAME COLOR
CALCIUM	ORANGE-YELLOW
COPPER	BLUE-GREEN
BARIUM	GREEN-YELLOW
SODIUM	YELLOW
LITHIUM	DEEP-PURPLE

2. Your teacher will conduct a flame test of various chemical substances. As each test is conducted, carefully observe any color(s) produced by the flame. Record this information beside the apporpriate solution letter in the data table provided.

SOLUTION	COLOR OF FLAME
A	
B	
C	
D	
E	

ANALYZE AND CONCLUDE

Write your answers on the lines provided.

1. Which solution, if any, contained barium ions?
_____

2. Which solution, if any, contained calcium ions?
_____

3. Which solution, if any, contained lithium ions?
_____

4. Which solution, if any, contained sodium ions?
_____

5. Which solution, if any, contained copper ions?
_____

6. Between each flame test, the loop must be dipped in a substance to be cleaned. Why is it important to clean the wire loop before testing for a given ion?

_____

_____

_____

Adriana Ocampo

Think WORK ACT

CRITICAL THINKING *Answer the following questions in complete sentences.*

1. What changes might occur if sunlight was unable to reach Earth's surface? Why might these changes affect Earth's animals?

   _____

   _____

   _____

2. How might the job of a planetary geologist differ from the job of a geologist studying Earth?

   _____

   _____

3. Why might scientists be interested in learning what elements are present in the atmosphere or soil of another planet?

   _____

   _____

GOING FURTHER
Complete three of the following.

BUILD YOUR PORTFOLIO

Use a dictionary or an encyclopedia to find the differences among meteors, meteorites, meteoroids, asteroids, and comets. Make a table of your findings that includes an illustration of each object.

USING COMMUNITY RESOURCES

In your community, find out if there is an astronomy club, a college or university group, or a planetarium that has sky-viewing activities. Make arrangements to attend one of these outings and report to the class on what you observe.

PERFECT YOUR SKILL

Using library resources make a drawing that shows the locations of the planets within the solar system as well as some of the known surface features of each planet. As an alternative, place the information you obtain in a table.

COOPERATIVE LEARNING

Use library resources or contact NASA using a computer to research some or the space probes that have been launched. Prepare a table or a report that indicates when the probes were launched, their destinations, and what information each probe is supposed to obtain.

RESEARCH AND REPORT

Research several hypotheses that explain the extinction of the dinosaurs. Summarize each hypothesis you research and explain which hypothesis you think is most likely.

ELLEN OCHOA

1958 -

1958 *Ellen Ochoa is* born in Los Angeles, California.

1985 *Ochoa* earns a Ph.D. in electrical engineering from Stanford University and applies for admission into the space program.

1990 *Ochoa* begins training at the Johnson Space Center of NASA as an astronaut.

1994 *Ochoa* goes into space again as the payload commander of a shuttle mission.

1980 *Ochoa* receives a B.S. in physics from San Diego State University.

1989 *Ochoa* is awarded the Hispanic Engineer National Achievement Award for Most Promising Engineer in Government.

1993 *Ochoa* becomes the first Latin American woman astronaut to make a space flight.

The conquest of space began long before people went to the moon. For some, it began with the telescopic observations of Galileo, the seventeenth-century Italian physicist and astronomer. For others, it began with the writings of Jules Verne, in his book, *From the Earth to the Moon*. Ellen Ochoa's conquest of space began in 1987, when she learned she had been selected by NASA to train for their astronaut program.

Ellen Ochoa was born in Los Angeles, California in 1958. She was one of five children in the family. While in junior high school, Ellen's parents divorced. Ellen then moved with her mother, Rosanne, to La Mesa, a suburb of San Diego. There, Ellen attended school and prepared for her future.

Ellen's mother was an inspiration and a role model for her daughter. In 1959, one year after Ellen's birth, Rosanne Ochoa began taking college courses. She spent the next 23 years working at her studies. Her hard work paid off in 1982 when she was awarded a bachelor's degree from San Diego State University. Rosanne Ochoa graduated

with a triple major in business, biology, and journalism.

Like her mother, Ellen believed in the value of education. While in junior high school, she did well in math and science. In 1975, Ellen graduated at the top of her class from Grossmont High School. She repeated this achievement when she graduated first in her class from San Diego State University with a B.S. degree physics in 1980.

After earning her bachelor's degree, Ellen went to Stanford University. She earned her masters degree in electrical engineering in 1981. Four years later, in 1985, she earned her Ph.D. in electrical engineering from the same school. While working on her Ph.D., Ochoa developed a method for using crystals to filter images. Ochoa later received a patent related to this work.

Ellen applied for admission into the NASA space program in 1985. That same year, she also began work as a research engineer at the Sandia National Laboratories in California. Ochoa remained at the labs for three years. While there he received two more patents in optical processing.

In 1988, Ochoa joined the NASA Ames Research Center in California. She was hired to lead a research group working on optical processes. Six months after taking this position, Ochoa was asked to become the Chief of the Intelligent Systems Technology Branch of the NASA/Ames Research Division.

In 1990, Ellen Ochoa was chosen by NASA to train as an astronaut. Her training began at the Johnson Space Center in 1990. After completing her training, Ochoa became the first Latin American female astronaut. Three years later, Ochoa became the first Latin American woman to go into space during a nine-day mission of the shuttle *Discovery*. Ochoa is one of only a few women to make history within the U.S. space program.

In her space flight of 1993, Ochoa served as a mission specialist. Her job was to use the shuttle's Remote Manipulator System to launch and then capture the *Spartan* satellite. The satellite was launched to gather data about the sun's corona. In 1994 Ochoa made a second shuttle trip, which lasted 12 days. On this trip, Ochoa served as the payload commander.

For her work as an engineer and an astronaut, Ochoa has received many awards. Among them is the Hispanic Engineer National Achievement Award for Most Promising Engineer in Government. She was presented with this award in 1989. In 1991, Ochoa was presented the Science Achievement Award by *Hispanic Magazine*. The next year, she received the Congressional Hispanic Caucus Medallion of Excellence. She has been further honored with the Women in Science and Engineering (WISE) Engineering Award, which she received in 1994.

Ellen Ochoa is well known for her public speaking appearances. She has stated that being a good role model is very important to her. She believes strongly in education and takes great pride in her accomplishments. Ochoa is hopeful that young Latin American children will see a bit of themselves in her.

Vocabulary

NASA is an acronym for the National Aeronautics and Space Administration, the government agency responsible for space exploration.

HISTORY MAKING WOMEN OF THE U.S. SPACE PROGRAM

1996: After spending 388 days aboard the Russian space station *Mir,* Shannon W. Lucid makes history as the American astronaut who has spent the longest time in space on a single mission.

1995: Eileen Collins makes history as the first woman astronaut to pilot a shuttle mission.

1993: In April, Ellen Ochoa makes history as the first Latin American woman astronaut to go into space. Ochoa makes her journey aboard a 9-day mission of the shuttle *Discovery*.

1992: On September 9, Dr. Mae Jemison becomes the first African American woman astronaut to go into space as a member of the *Endeavor* crew.

1986: On January 28, the space shuttle *Challenger* explodes 23 seconds after take-off. All seven crew members are killed, including astronaut Judith A. Resnick and teacher Crista McAulliffe, the first civilian to take part in the shuttle program.

1983: On June 18, Sally K. Ride becomes the first American woman to go into space on a flight of the *Challenger.*

1981: On April 12, the space shuttle program begins with the launch of *Columbia.*

It's YOUR TURN

COMPARING IMAGES FORMED BY LENSES

MATERIALS (per group of two)

convex lens, concave lens, paper, pencil

BACKGROUND INFORMATION

A *lens* is a curved, transparent material that freely allows light to pass through it to form an image. There are two main types of lenses: convex lenses and concave lenses. A convex lens is one that is thicker at its center than at its edges. A concave lens is thinner at its center than at its edges. Each type of lens forms a different type of image.

PROCEDURE

1. Examine each lens you have been given. Make a sketch showing the shape of each lens. Using the illustrations and the Background Information, identify each lens in your sketch as concave or convex.

2. Using the convex lens, view the Background Information text. Move the lens toward and away from the page until a clear image forms. Note if and how the image changes as you move the lens.

3. Repeat step 2 using the concave lens.

Convex Lens   **Concave Lens**

ANALYZE AND CONCLUDE

Answer the following questions on the lines provided.

1. Describe how the shape of the convex lens differs from that of the concave lens.

_____

_____

2. Describe the image formed by the convex lens.

_____

_____

3. Describe the image formed by the concave lens.

_____

_____

4. Based upon your observations, what kinds of instruments might use convex lenses?

_____

_____

_____

5. What kinds of instruments might use concave lenses?

_____

_____

_____

Name _____   Date _____

Think WORK ACT

CRITICAL THINKING *Answer the following questions in complete sentences.*

1. How might a degree in engineering be helpful to an astronaut?

   _____

   _____

2. Ellen Ochoa has stated that being a role model is one of her top priorities. What traits of Ochoa help her to serve as a role model?

   _____

   _____

   _____

3. Ellen Ochoa's mother began college the year after Ellen was born. She received her bachelor's degree 23 years later. What impact do you think this had on Ellen as a student?

   _____

   _____

GOING FURTHER   *Complete three of the following.*

BUILD YOUR PORTFOLIO

Obtain a copy of the government publication, *Spinoff*. Create an illustrated table or diagram that identifies five products and processes that have resulted from space travel. For each spinoff, identify its use in space as well as its use or uses on Earth.

USING COMMUNITY RESOURCES

Write to or e-mail NASA to get information about women in the space program. Use the information you obtain to create a scrapbook. Display the letter you sent to NASA and the information you placed in your scrapbook where it can be studied by other students.

JOURNAL WRITING

In your journal, describe ways in which becoming an astronaut interests you. What skills do you think you would need to improve on to become an astronaut? What skills do you have that might be well suited to a career as an astronaut?

COOPERATIVE LEARNING

Use library resources to find out what is meant by the term *weightlessness* when speaking about space travel and how this condition affects people aboard a space vehicle. Taking the effects of weightlessness into account, prepare and design packaging for three foods you would take into space.

RESEARCH AND REPORT

Read Jules Verne's *From the Earth to the Moon*. Write a brief summary of the novel. Describe one way in which the novel accurately depicts space travel and one way in which the description of space travel is completely fictional.

HERMELINDA
RENTERIA

1960 -

1960 *Hermelinda Renteria* is born in Los Llamas Zacatecas, Mexico

1983 *Renteria* receives her B.S. in engineering from the Universidad Autonome de Guadalajara in Mexico.

1988 *Renteria* serves her first of four terms as president of the San Francisco chapter of the Society of Hispanic Professional Engineers (SHPE)

1963 *The Renteria* family moves to the United States.

1984 *Renteria* begins her career with the Pacific Gas and Electric Company in Ventura California.

1989 *Renteria* serves as national secretary of SHPE.

In 1951, The National Society of Professional Engineers (NSPE) founded National Engineer's Week. Each February, engineers throughout the country set up displays to make people aware of the work they do. Many engineers visit schools and colleges to speak with students about careers in engineering. Hermelinda Renteria is a construction engineer. She tries to stimulate interest in science, math, and engineering every week of the year. Renteria, who is a Mexican American, is especially interested in serving as a role model for Latino women.

Hermelinda Renteria was born in Los Llamas, Zacatecas, Mexico in 1960. She was one of four children of Maria and Santiago Renteria, who were farm workers.

In 1963, the Renteria's moved their family to Chico, California. During the next three years, the family moved often in search of farm work. During this three-year period, the family often made their home in housing shared with other families in farm labor camps.

In 1966, the Renteria family settled in Watsonville, California. Here, the children received their early education in public schools. Later, when Hermelinda and her sister were of high school age, their father insisted that the girls be sent to a private Catholic School.

After high school, Hermelinda returned to Mexico to attend college. She hoped that living in Mexico would help her learn more about her roots. In Mexico, Renteria enrolled as an engineering student at the Universidad Autonome de Guadalajara in 1979. She received her B.S. in engineering from the school in 1983. Renteria was one of only three female students to graduate with a B.S. degree.

After college, Renteria remained in Mexico. She took a short-term job as an engineering aide with the Jalisco State Department of Public Work in Guadalajara. In 1984, Renteria returned to California when she learned that her parent's business was failing. She worked with her family

Hermelinda Renteria

to save the business. Later that year, when the business began to profit, Renteria returned to her engineering career. She went to Ventura, where she worked as an engineering aide and draftsperson.

During 1984, Renteria held several different jobs. She finally accepted a job as a field engineer with the Pacific Gas and Electric Company (PG&E). Since beginning work for PG&E, Renteria has held several positions within the company. Among these were her work as a field engineer at the Diablo Canyon Nuclear Power Plant in San Luis Obispo and as an assistant construction superintendent for the Diablo Canyon Power Plant in San Francisco. While with PG&E, Renteria also re-designed several departments inside of the PG&E building near San Francisco. For her work on this last project, she received a Performance Recognition Award from the company in 1988.

In 1992, Renteria took a job with PG&Es San Francisco Bay Power Plant. Here, Renteria works as the contracts and technical services supervisor. This job allows Renteria to work with outside contractors on PG&E projects. In this role, she gets to work in the field, an opportunity she very much enjoys. The contracts and technical services supervisor's job is to make sure the contractors do their work according to PG&E standards.

Renteria has stated that being Latin American has not been a problem in her field. However, being a woman has been a problem. Most of her engineering duties, including those as a field engineer, involved doing paperwork. Renteria had to request work outside the office. She finally got

VOCABULARY

DRAFTSPERSONS are the people who prepare the plans and drawings for construction projects.

her chance and proved herself worthy.

In addition to her engineering work, Renteria tries to be involved in her community. Much of this work involves encouraging women's interest in the sciences and engineering. To help in this area, Renteria participates in several mentoring programs. In addition, Renteria has been on the board of directors for the San Francisco Bay Area Girl Scouts.

For her work with young people and services to the community, Renteria received the Anti-Defamation League B'nai B'rith, "Woman on the Move" Certificate of Honor in 1989. In 1990, the San Francisco YWCA presented her with a Certificate of Merit, and in 1991 Skyline College of San Bruno, California awarded her with a Certificate of Appreciation.

Renteria is also active in professional organizations. She has been an active member in the Society of Hispanic Professional Engineers (SHPE). Between 1988 and 1993, she served as president of the San Francisco chapter of SHPE. From 1989 through 1991, she was the national secretary of SHPE.

For her professional work and dedication to her field, Renteria has received many awards. In 1988 she was granted an award by SHPE. The same year, she was recognized by the Mexican Association of Mechanical and Electrical Engineers for her contribution in organizing that group's Third International Conference on Engineering and Technology. In 1991, she was recognized again by SHPE for her "Outstanding Leadership and Dedication" to her field.

It's YOURTURN

CONSTRUCTING A SCALE DRAWING

MATERIALS (per student)

freshly sharpened pencil, paper, tape measure, 12-inch ruler

BACKGROUND INFORMATION

A structure at a construction site is often too large to be drawn on paper as its actual size. When drawing plans for such a structure, a draftsperson must reduce the sizes of all measurements involved in the structure without changing the overall dimensions. To do this, a scale is developed in which the drawn size is compared to the actual size. For example, if a blueprint states that 1/4" inch equals 1 foot, then each 1/4 inch on the paper represents 1 foot on the structure. Such a scale would be recorded as 1/4 in. = 1 ft.

PROCEDURE

1. Assign labels to the walls of your classroom such as Front, Back, etc.

2. Measure each wall of your classroom using your tape measure. Record the measurements. After the overall size of each side of the room is taken, identify the locations and sizes of any doors and windows that are located along each wall.

3. On a sheet of paper, make a drawing that represents a "bird's eye" view of your classroom (See sample drawing). Make 1/4 inch on your drawing equal to 1 foot.

4. After you draw the general shape of the classroom, add the locations of doors and windows along their appropriate walls. Be sure to draw these structures to scale.

5. Measure your teacher's desk and its location from the two nearest walls.

6. While maintaining your scale, add the desk to your classroom drawing.

ANALYZE AND CONCLUDE

Answer the following questions on the lines provided.

1. What was the scale of your drawing?

   _____

2. Why should the scale appear on the drawing?

   _____

   _____

3. Why is it important to use a sharp pencil when making construction drawings?

   _____

   _____

4. Why are scale drawings useful?

   _____

   _____

Sample Scale Drawing

Window

Scale

1/4" = 1'

Door

Hermelinda Renteria

Think WORK ACT

CRITICAL THINKING *Answer the following questions in complete sentences.*

1. List ways in which Hermalinda Renteria was discriminated against because of her gender.

   _____

   _____

2. Do you think women are still discriminated against in some fields? Explain.

   _____

   _____

3. Of what importance are blueprint drawings?

   _____

   _____

GOING FURTHER

Complete three of the following.

BUILD YOUR PORTFOLIO

Find examples of different things that make use of scales. Identify the scale used for each item you identify. Write a brief description of how the use of a scale was important to each item. Include your descriptions in your portfolio.

ALTERNATE ASSESSMENT

Make a scale drawing of the floor plan of your home. On the drawing, indicate two escape routes from each room of the house that could be used in an emergency.

JOURNAL WRITING

Obtain blueprints of buildings, equipment, or properties. Write a detailed description of what each blueprint represents.

COOPERATIVE LEARNING

In your group, create a table listing several branches of engineering and the responsibilities of a person working in each. **Hint:** Start with the branches of engineering in this book work.

RESEARCH AND REPORT

Renteria was not the only woman who met with discrimination because of her gender. Do research to find the names of other women who faced similar problems while developing their careers.

RODRIGUEZ-TRIAS

1929-

1929 *Helen Rodriguez* is born in New York City.

1960 *Rodriguez* graduates first in her class from the University of Puerto Rico Medical School.

1975 *Rodriguez* joins the faculty of the Biomedical Program at City College in New Yor

1957 *Rodriguez* receives her B.S. from the University of California at Berkeley.

1970 *Rodriguez* becomes director of the Department of Pediatrics at the Lincoln Hospital in New York.

1990 *Rodriguez* resigns from her position with the Department of Health in New York.

Children are often high on the list of people who are in need of medical services. Yet, for a variety of reasons, they are often overlooked. Among those who have worked to improve the quality of pediatric health care is Dr. Helen Rodriguez-Trias, a Latin American physician.

Helen Rodriguez-Trias was born in New York City in 1929. Soon after, the Rodriguez family moved with their daughter to their homeland of Puerto Rico. The family lived on a coffee plantation owned by Helen's grandparents.

Helen Rodriguez spent her early childhood in Puerto Rico. Her father made his living by selling men's clothing. Helen's mother was a teacher. Helen received her early elementary education in a private Catholic school.

When Helen Rodriguez was 10, her parents divorced. Helen moved with her mother back to New York City. Helen's mother was a strong and independent woman who valued education. She served as a role model to her daughter and was the wage earner for the family.

While living in the city, Rodriguez attended George Washington High School. She was a good student and did well in the sciences. At a young age, Helen knew that she would somehow make science a part of her life's work. After graduating from high school, at the urging of her motherRodriguez returned to Puert Rico to attend college, While in college, Helen developed a strong interest in politics. Soon this intere grew so strong that Helen left colle to take an active role in political campaigns. Helen returned to the United States and began to work on the political campaign of Henry Wallace.

Soon, after returning to the state Rodriguez married. During this marriage, she had three children. She also went to college. In 1957, she received a B.S. degree from the University of California at Berkeley.

After completing college, Rodriguez decided s was not happy with the direction her life was taking. She traveled with her three children to Puerto Rico on what she called a "vacation." Onc back in Puerto Rico, Helen decided to remain there and divorce her husband.

While living in Puerto Rico, Helen Rodriguez decided to study medicine. She enrolled in the School of Medicine at the University of Puerto Rico. In 1960, Rodriguez graduated from medical school at the top of her class.

While working on her medical studies, Rodriguez remarried. Her new husband took over the raising of the children while Helen worked on her residency. During her residency, Helen specialized in pediatrics and neonatology. During this period, she also had another child.

Over time, the demands of a career in medicine took their toll on Helen's marriage. After 16 years of marriage to her second husband, Rodriguez again divorced. Following her divorce in 1970, she again returned to New York.

In 1970, Rodriguez became director of pediatrics at the Lincoln Hospital in the South Bronx. Four years later, she became associate director of pediatrics for primary care at St. Luke's-Roosevelt Hospital Center. She was also served the director of Roosevelt Hospital's Children and Youth Program.

The communities served by both the Lincoln and St. Luke's-Roosevelt hospitals were largely Latino. While working at both hospitals, Rodriguez spent much of her time trying to get hospital employees to become aware of the individual and cultural needs of the people in the community. She was largely unsuccessful in these goals. In addition, Rodriguez herself often faced discrimination because of her race and gender at each of these facilities.

In 1975, Rodriguez joined the staff of the Biomedical Program at City College in New York. While there, she taught a course in social medicine. The course stressed the need for physicians to become involved in and aware of the individual and cultural needs of the members of the communities they served. To help her students become aware of these needs, Rodriguez required them to go into the community to speak with people and find out what medical services they needed.

Following her work in the Biomedical Program at City College in New York, Rodriguez continued to teach. In the early 1980s, she taught classes at Fordham University, the Sophie Davis Center of City College, and at the Columbia College of Physicians and Surgeons. Later, she worked for the New York City Department of Health. She left this position in 1990 to spend more time with her grandchildren.

In addition to her work in medicine, Rodriguez works to educate others about women's and children's rights. She is active in many groups concerned with these issues. Some of these include the American Board of Pediatrics, the National Resource Center for Children in Poverty, the Board of American Public Health Association, and the Center for Constitutional Rights. Most recently, Rodriguez has served as president of the American Public Health Association and as a member of the National Center for Children in Poverty's Council of Advisors.

> ## Vocabulary
>
> NEONATAL MEDICINE involves the study and treatment of the medical needs of newborns.
>
> PEDIACTRICS is the branch of medicine that deals with taking care of the health of children

It's YOURTURN

Hands-On Activity

ANALYSIS OF MORTALITY RATES
OF CERTAIN INFECTIOUS DISEASES

BACKGROUND INFORMATION

It is important that children receive inoculations against certain infectious diseases. Most school systems in the United States require children to get certain vaccines before they can be admitted to school. Hospitals and community health centers often provide programs that help educate people about such matters. Mortality tables show the number of people who die from certain types of diseases. The data in such tables illustrate the importance and success of vaccination programs.

PROCEDURE

1. Study the table. Use the information in the table to answer the Analyze and Conclude questions.

Death Rates for Selected Diseases (per 1,000,000 people)								
Disease	1900	1920	1945	1950	1980	1990	1993	1994
Typhoid fever	267	73	2	1	0	0	n/a	n/a
Measles 100	73	6	3	0	0	0	0	
Scarlet fever	118	4	1	2	0	0	n/a	n/a
Whooping cough	107	89	10	7	0	0	0	0
Diphtheria	327	137	7	3	0	0	n/a	n/a
Tuberculosis (TB)	1847	967	333	225	233	328	309	314
Pneumonia & Influenza	1843	1403	413	313	233	328	309	314
Cardiovascular disease	3595	3699	4931	5108	4345	3683	3647	3603
Cancers	677	869	1340	1398	1825	2032	2060	2066

ANALYZE AND CONCLUDE

Answer the following questions on the lines provided.

1. Which diseases show the greatest decline in the number of deaths for the years shown? What might be some reasons for the decline?

_____

_____

2. Which diseases show an increase in the number of deaths over time? What might be some reasons for the increases?

_____

_____

3. For which diseases do there not seem to be large increases or declines over time?

_____

_____

4. During what years are the greatest drop in deaths shown for typhoid fever; measles; scarlet fever; whooping cough; diphtheria; tuberculosis? What do you think accounted for this change?

_____

_____

Think WORK ACT

CRITICAL THINKING *Answer the following questions in complete sentences.*

1. What evidence supports the work of Helen Rodriguez in the establishment of educational programs at various community centers?

   _____

   _____

2. Why are today's children who have not been vaccinated against certain diseases in greater danger of getting infectious diseases than those in earlier years?

   _____

   _____

3. Why might infectious diseases occur with greater frequency in developing countries than in industrialized nations?

   _____

   _____

GOING FURTHER
Complete three of the following.

BUILD YOUR PORTFOLIO

Make a table that lists five infectious diseases, how they transmitted, and what methods of treatment or preventions are available for preventing each disease.

USING COMMUNITY RESOURCES

Interview your family doctor, the school nurse, or another person in the healt- care profession to find out what diseases young children should be vaccinated against and at what ages and intervals such vaccines should be given. Use the information you obtain to create a table.

JOURNAL WRITING

There is controversy over whether a child with chicken pox should go to school or stay at home until no longer contagious. Write a short essay on your opinion of this controversy.

COOPERATIVE LEARNING

Have each group member research a different infectious disease. For each disease, find out its cause, its symptoms, how it is treated, and whether a vaccine is available to help prevent the spread of the disease. Combine all data collected in a group report.

RESEARCH AND REPORT

Research the location and function of health services in your community. Present this as a report that could be distributed to interested persons in your area.

CAROL SANCHEZ

1961 -

1961 *Carol Sanchez* in born in Tucson, Arizona.

1985 *Sanchez* begins her career with Hughes Aircraft.

1990 *Hispanic Magazine* gives *Sanche* its Most Promising Engineer Award.

1984 *Sanchez* receives her B.S. in systems engineering from the University of Arizona.

1987 *Sanchez* is awarded the Superior Performance Award by her company. The award is the highest that can be granted to an employee.

You may have read about guided missiles or watched one fired in an action movie. A missile is any object that is launched at a target. A guided missile is one that has some type of system to direct and control its flight. One outstanding woman doing research on missiles is Latin American engineer Carol A. Sanchez.

Carol Sanchez was born in Tucson, Arizona, on April 24, 1961. Both her parents and grandparents were also born in the United States. However, her great grandparents were born in Mexico. Thus, Carol and her siblings are third generation Mexican Americans.

Education was always considered important to the Sanchez family. Because of this, the children rarely missed a day of school. Carol Sanchez enjoyed elementary school, especially class trips to local museums. As a result of these trips, she became very interested in the history of Native Americans and Latin peoples and their influence upon American culture.

In high school, Sanchez took typing, shorthand, and other subjects that would prepar her for a secretarial career. She avoided science and mathematics. At the time, it wa common to channel girls away from technical subjects and into commercial subjects. Girls were expected to enter the business world as secretaries and bookkeepers. However, in the 1970s, Carol Sanche attended a career fair. While attending the fair, she became intrigued by computers. This interes in computers caused her to change her career plans.

After graduating from high school in 1980, Carol Sanchez enrolled as a student at the University of Arizona. The liberal arts and business programs offered by the school did not appeal to Sanchez. She was more interested in computer sciences. Engineering wa as close as she could get to this field. Thus, she chose engineering as her field of study. Four yea later, in 1984, Carol Sanchez graduated with a B. in systems engineering.

INDUSTRIAL ENGINEER

In 1985, Carol Sanchez began her engineering career with the Hughes Aircraft Company. While with the company, Sanchez has worked mostly in missile research and design. In her work, Sanchez researches the use of electronics and computers in missile guidance systems. The systems she develops use TV, infrared, and lasers.

All missile guidance systems have importance to the military. However, there are also many peaceful applications of missiles and the technology used to guide them. For example, TV-missile guidance systems make use of a television camera mounted in the nose of the missile. This camera allows an operator at another location to see where the missile is going and change its course if necessary. In July 1996, an U.S. passenger plane crashed into the ocean off the coast of Long Island, New York. Parts of the plane were scattered along the ocean floor, often at depths of more than 100 feet. Using the technology created for TV-guided missiles, the U.S. Navy launched an underwater probe. The camera mounted in the probe helped locate pieces of the wreckage. Divers and salvage crews could then bring the materials to the surface.

Rocket engines are usually used to power guided missiles. The same types of rocket engines have been used for peaceful purposes such as powering space vehicles, space probes, and to launch artificial satellites. The working of these engines is based on Newton's Third Law of Motion.

Carol Sanchez's main job is to study the processes and machinery used in developing and producing medium-range air-to-air missiles. However, Sanchez has also used her skills to study the work patterns of employees involved in the program. Her studies of how people work led to some changes at the company. The changes, in turn, led to greater production output and estimated savings of almost $1 million at the company.

In recognition of her accomplishments, Hughes Aircraft has presented Sanchez with several awards. In 1987, only two years after beginning her work with the Company, Sanchez was given the Superior Performance Award. This is the highest award an employee of the company can receive. Two years later, Sanchez was recognized by the company again with two High Performance Team Awards.

Carol Sanchez does not limit her talents to her work. She has done much in the community to encourage others to become involved in science and engineering. She is a yearly participant in the University of Arizona preengineering workshops. She is a frequent speaker at the Youth Convention of the League of United Latin American Citizens. Sanchez also takes part in high school and elementary school programs directed toward women and minorities.

> **Vocabulary**
>
> NEWTON'S THIRD LAW OF MOTION states that for every action there is a reaction that is equal in force, but opposite in direction.

For her work in the community as a mentor and role model, Carol Sanchez has been granted much praise. In 1990, *Hispanic Engineer* magazine awarded her the Most Promising Engineer Award. The same year, the magazine *Professional* recognized her as one of the top 20 minority engineers in the country. One year later, in 1991, Sanchez was featured as a role model in an exhibit honoring Latino achievers that was produced by the AT&T corporation.

In addition, Sanchez is a member of several professional organizations. Among them are the Hughes Hispanic Employees Association, the Society of Hispanic Professional Engineers (SHPE), and the American Institute of Industrial Engineers. She has also launched a mentoring program to help Latino students at a local Tucson, Arizona high school. Sanchez is married to Michael W. Conrad, an industrial engineer at Hughes Aircraft.

It's YOUR TURN

THE FORCE THAT PROPELS THE MISSILE

MATERIALS (per pair of students)

2 (3" x 5") index cards, 2 rubber bands, 2 paper clips, balloon, stapler

BACKGROUND INFORMATION

As someone steps forward from a rowboat onto a dock, the boat moves away from the dock. The movement of the boat is explained by Sir Isaac Newton's Third Law of Motion. This law states that for every action, there is a reaction of equal force in an opposite direction. In the case above, the person jumping onto the dock performs the action while the boat, which moves in an opposite direction, produces the reaction. This principle explains how a missile gets the force needed to travel through the air.

PROCEDURE

1. Measure 1/2" down from the center of the 3" side of the index card. Staple the rubber band to the card at this point. Attach a paper clip to the rubber band.

2. From the other side of the card, measure 1/8" divisions for 1 1/2" as shown in the diagram.

3. Repeat steps 1 and 2 with the other index card.

4. Connect the two paper clips together, and place the cards on your desk.

5. Move one card while you hold the other in place. Note the positions of the clips on both cards as you move the card.

6. Inflate a balloon, holding the neck closed to prevent air from escaping. Release the balloon and observe what happens.

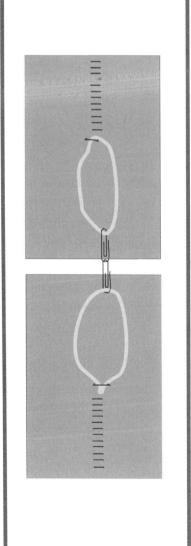

ANALYZE AND CONCLUDE

Answer the following questions on a sheet of paper.

1. In steps 5 through 6, what was the action? What was the reaction?

   _____
   _____
   _____

2. How does the activity with the rubb bands illustrate Newton's Third Law of Motion?

   _____
   _____
   _____

3. How does this activity illustrate how rocket engine works?

   _____
   _____
   _____

Carol Sanchez

Think WORK ACT

CRITICAL THINKING *Answer the following questions in complete sentences.*

1. How might the ideas used in the TV guidance system be used in medicine?

   _____

   _____

   _____

2. Use Newton's third law of motion to explain why it usually takes more than one firefighter to hold a hose when water is coming out at full force.

   _____

   _____

   _____

3. What are some uses of missiles in space exploration?

   _____

   _____

GOING FURTHER
Complete three of the following.

BUILD YOUR PORTFOLIO

Do research to find out about multistage rockets. Create an illustration that shows and labels different parts of a multistage rocket.

COOPERATIVE LEARNING

Work as a team to research the Hero of Alexandria's turbine steam jet engine and how turbogenerators are used to create electricity from wind. Write a report that explains how Newton's Third Law of Motion applies to each of these devices.

RESEARCH AND REPORT

Do research about some of the space missions carried out by NASA. Construct a chart listing the names of the rocket used for each mission, where the rocket was sent, and the year of the mission.

JOURNAL WRITING

Governments spend a great deal of money on rocket and missile research. In your journal, explain your feelings about the spending of money on such research.

Alvarez, Anne Maino (1941 -) Maino is a plant pathologist who works at the University of Hawaii. She has done much work on bacterial diseases in fruits and crops in Argentina, Costa Rica, and Mexico. A specific area of interest in her work was research into diseases of the papaya.

Alvarez, Sara Lynn (1953 -) Sara Lynn Alvarez is a researcher in the field of optometry and an assistant professor of optometry at the University of Alabama. She was a bubble chamber technologist, has a Ph.D. in optometry, and has produced many publications pertaining to problems of the eyes.

Alvariño, de Leira, Angeles (1916 -) Alvariño is a biochemist, a Fulbright scholar, and a discoverer of 11 new species of organisms. She Alvariño has worked as a marine biologist and as an oceanographer at the Spanish Institute of Oceanography (1952), researched zooplankton, and was awarded Great Silver Medal of Galicia by King Juan Carlos I and Queen Sophia of Spain.

Asunción, Elena Charola (1942 -) Asunción is an analytical and physical chemist who is doing research in solid-state chemistry, polymorphism, and X-ray crystallography.

Bernárdez, Teresa (1931 -) Bernárdez is a psychiatrist concerned with problems faced by women. She has numerous publications on this subject. Among her achievements are her receipt of the American Medical Association Physician's Recognition Award in 1970, the Menninger School of Psychiatry Teacher's Recognition Award in 1971, the Pawlowski Foundation Peace Award in 1974, and the Michigan State University Faculty Women's Association Distinguished Faculty Award in 1982.

Borras, Caridad (1942 -) Caridad Borras is a radiological physicist for the West Coast Cancer Foundations. She is a clinical assistant professor at the University of California and does research in the radioembryo pathological effects of high LET nuclides. Borras has been involved in the study of the physics of radiology and radiation therapy.

Candeles, Graciella (1922 -) Candeles is a cytologist and professor of cell and molecular biology at the University of Puerto Rico she is a a visiting professor at the University of Syracuse the Rockefeller Institute, the City University of New York (CUNY), and the Medical College of Georgia. She received a special science award from the Puerto Rico Institute in New York in 1985.

Faustina, Solis (1923 -) Faustina works for the public health administration. She was the first full professor at the University of California at San Diego School of Medicine who did not hold medical degree or a doctorate. Faustina has developed a program for agricultural migrant workers that provides them with education, health care, day care for their children, and housing.

Garcia, Catalina Esperanza (1941) Garcia is Latin American anesthesiologist. She was the first Latina to graduate from the University of Texas Medical School. She was also a government appointee to the Texas State Board of Medical Examiners and was involved in the Mexican American Business and Professional Woman's Foundation. Throughout her career, Garcia has served as role model for school children, especially those of Latino backgrounds.

Ginorio, Beatriz Angela (1947-) Ginoria was first Puerto Rican to direct a center for research on women in the United States. She has produced publications for the American Psychological Association and the Mexican American Women National Association. In addition, Ginorio has taught at the University of Puerto Rico, the University of Illinois, Bowling Green State University, and the University of Washington.

Gonzalez, Paula (1932 -) Paula Gonzalez is a Latin American scientist focusing on the areas of bioethics and environmental science. Gonzalez studied cell physiology in addition to environmental science. She is chairman of the Biology Department at the College of Mt. St. Joseph and has conducted research in cellular activity and cell environments.

gnacio, Carmella (1942 -) Carmella Ignacio is recognized as the first Latina registered nurse at the Grossmont Hospital in San Diego, California to reach Level IV for clinical nurses. Ignacio was born to parents of mixed heritage both Native American and Latino. Her mother was of the Chico family and her father of the Pueblo. As a child, she attended the Topowa Elementary School and the St. John's Indian School in Arizona. She later went to the University of Arizona and the St. Mary's School of Nursing, from which she received her nursing degree. She has provided medical service to thousands of Native Americans at the Indian Health Center and has worked with both state and federal agencies to improve health care to Native Americans living in urban areas throughout California.

Mercado, Teresa (1921 -) Mercado is a specialist in histochemistry and cell pathology. She has worked as a research physiologist at the National Institute for Allergy and Infectious Diseases. She has also conducted studies on parasitic diseases for the National Institutes of Health. Specifically, she has done extensive work on malaria and trypanosomiases.

Urdaneta, Maria Luisa (1931 -) Urdaneta is a nurse, psychologist, sociologist, and anthropologist. She has been employed as a research associate at the University of Texas Health Science Center in San Antonio and as a professor at the University of Texas. She was recognized by the San Antonio Women's Hall of Fame, the Texas Diabetes Council, the National Institutes of Health, the Mexican-American Business and professional Women's Club of San Antonio, and the National Chicano Research Network.

Villa-Komaroff, Lydia (1947 -) Lydia Villa-Komaroff is a Latin American scientist who works in the area of molecular biology. Villa-Kamaroff focuses most of her attention to the field of neurology and the functioning of the nervous system. In this role, she is working on the analysis of a protein that may be the cause of a rare condition called megalencephaly. This condition results in abnormal growth of the fetal brain, resulting in a brain that is larger than normal. In addition to her research, Villa-Komaroff served as an associate professor of neurology at the Harvard Medical School before taking an administrative position with Northwestern University.

The science classroom is a place with potentially dangerous situations. However, the classroom can be a safe place to work if you work carefully. The following safety rules will help to protect you and your classmates from potential dangers in the science classroom. Read all of these guidelines before conducting any science activity. In addition, as you prepare to carry out a specific activity, review the guidelines that are given on the activity page.

Safety Guidelines to Follow Before the Activity

1. Make sure your teacher or another adult is in the classroom to supervise your work before carying out any activity.

2. Completely read the activity safety notes and procedure before beginning an activity. Make sure you gather or know the location of any safety equipment that may be required if an accident does happen.

3. Know where all safety equipment is stored and how to use the equipment. Safety equipment should include a fire blanket, dry chemical fire estinguisher, eye wash, fume hood, safety goggles and laboratory aprons, and a first-aid kit.

4. Know how to evacuate your classroom and your school should an emergency occur. Also learn the location of the nearest telelphone in case you need to call for help.

5. Inspect all glassware that is to be used in the activity for cracks or chips. Do not use glassware having cracks or chips. Report broken glasssware to your teacher and dispose of the glassware as instructed.

Safety Guidelines to Follow During an Activity

1. Never run in the classroom or play games during an activity.

2. Wear safety goggles when working with chemicals, open flames, or when instructed by your teacher. Wear an apron or laboratory coat to protect your clothing.

3. Wash and dry your hands before and after each activity.

4. Wash all glassware and equipment thoroughly before and after it is used.

5. Never eat or drink in the science classroom.

6. Never taste any substance or draw any material into your mouth.

7. Do not directly inhale chemical substances. If you must observe the odor of a substance do so only when directed and by using your hand to waft the fumes toward your nose.

8. Keep your work area clean and free of unneeded items.

9. Report any accident that occurs to your teacher immediately. If you spill any substance, wash it off with water and report the spill to your teacher.

10. Keep all materials away from open flames. Tie back long hair and secure loose clothing when working with open flames.

11. If something does catch fire in the classroom or if your clothing catches fire, DO NOT RUN. Smother the fire with the fire blanket or get under a safety shower.

12. Use caution when working with scissors, knives, or other sharp objects. Use these objects to cut in a direction away from your body.

13. When heating a test tube, always point the mouth of the test tube away from yourself and others.

14. Never mix chemicals together unless instructed to do so. Many chemicals can be explosive or dangerous when combined with other chemicals.

15. If you break glassware, do not pick up the broken glass with your hands. Sweep broken glass with a broom and dispose of it in a container labeled for glass disposal.

Safety Guidelines to Follow After the Activity

1. Do not return unused chemicals to their original containers. Follow the instructions given by your teacher for the appropriate way to dispose of used chemicals.

2. Clean up your work area. Make sure all materials are washed and put away in the area designated by your teacher.

3. Check to make sure all Bunsen burners, gas outlets, and water faucets are turned off before leaving the classroom.

4. Wash and dry your hands after conducting an activity.

GLOSSARY

AIDS an acronym that stands for acquired immune deficiency syndrome, a group of illnesses that are caused by a viral attack on the immune system

astronomy the study of stars and celestial objects

autoimmune disease results when the body attacks its own cells as if they were agents of disease

binary star a star system made up of two stars that revolve around each other

botanist a scientist who studies plants

civil engineer a person trained in the design and building of public works

colon also known as the large intestine; the part of the digestive system that is responsible for absorbing fluids before the remaining products from food are eliminated from the body as waste

dermatology the branch of medicine that deals with the health of the skin.

draftspersons the people who prepare the plans and drawings for construction projects

ecology the branch of science that studies the relationships between living things and their surroundings

extinction the dying out of a type of organism

genus a group of several species that share many characteristics

geochemist a scientist who studies the chemical makeup of Earth or of other bodies in space

groundwater the water that collects below the soil's surface in rock and soil layers; often the source of water for a community

hazardous materials substances that pose a threat to human health and the environment because they are poisonous, corrosive, flammable, explosive, or radioactive

herbarium a collection of dried plants

HIV an acronym that stands for human immunodeficiency virus, the virus that causes AIDS

immunochemist a scientist who studies the chemical factors involved with the body's resistance to disease

immunology branch of medicine that deals with studying the body's resistance to or ability to fight disease

liaison a person who works to coordinate activities or to bring people together to work on a common goal

meteorite a particle of matter from space that reaches Earth

NASA an acronym for the National Aeronautics and Space Administration, the government agency responsible for space exploration

neonatal medicine involves the study and treatment of the medical needs of newborns

Newton's Third Law of Motion states that for every action there is a reaction that is equal in force but opposite in direction

ornithology the branch of biology dealing with the study of birds

paleontologist a scientist who studies ancient life forms from their fossils

pediatrics the branch of medicine that deals with taking care of the health of children

planetary geologist a scientist who studies the physical features of planets

psychology the branch of science that studies behavior and mental processes

physiology a study of how the cells, tissues, and organs of a body function

spectra the abnd of colors formed when light is passed through a prism

systems analyst a person who studies the procedures of an activity to define its goals and to make sure that the activity is being done efficiently

tropical diseases illnesses that generally occur from exposure to disease-causing organisms that thrive in only tropical environments

immunology branch of medicine that deals with studying the body's resistance to or ability to fight disease

visa a permit that allows a person to travel in a certain foreign country for a specific period of time

water purification plant treatment facility that processes dirty water to make it clean for human use.

INDEX

A

activities, 6-7, 10-11, 14-15, 18-19, 22-23, 26-27,
    30-31, 34-35, 38-39, 42-43, 46-47, 50-51,
    54-55, 58-59, 62-63

Alternate Assessment, 23, 27, 31, 55

Analysis of Mortality Rates of Certain Infectious
Diseases, 58

Analyzing and Concluding, 6, 10, 14, 18, 22, 26,
    30, 34, 38, 42, 46, 50, 54, 58, 62

Analyzing Drug Abuse Death Rates, 10

Analyzing the Effects of Alcohol in the Blood, 42

Analyzing Trends in Home-Computer Use, 18

Build Your Portfolio, 11, 19, 23, 27, 31, 35, 39,
    43, 47, 51, 55, 59. 63

Cleaning Dirty Water, 6

Comparing Images Formed by Lenses, 50

Concept Mapping, 7, 11, 14, 26, 43

Constructing a Scale Drawing, 54

Cooperative Learning, 7, 11, 14, 19, 23, 26, 31,
    35, 39, 43, 47, 51, 55, 59, 63

Critical Thinking, 7, 11, 15, 19, 23, 27, 31, 35,
    39, 43, 47, 51, 55, 59, 63

Evaluating Public Service Messages, 26

Force that Propels the Missile, the, 62

Identifying Elements Using a Flame Test, 46

Identifying Trees, 38

Journal Activity, 23, 27

Journal Writing, 7, 14, 19, 35, 39, 43, 51, 55,
    59, 63

Leaching and Groundwater, 30

Perfecting Your Skill, 14, 35, 47

Researching and Reporting, 7, 11, 14, 19, 23, 27,
    31, 35, 39, 43, 47, 51, 55, 59, 63

Social Insect Behavior, 22

Spectral Analysis of Light, 34

Tropical Diseases and Their Transmission, 14

Using Community Resources, 7, 11, 19, 31, 47,
    51, 59

Alvarez, Anne Maino, 64

Alvarez, Sara Lynn, 65

Alvarino, de Leira, Angeles, 64

Asuncion, Elena Charola, 64

B

Bernardez, Teresa, 64

Borras, Caridad, 64

C

Candeles, Graciella, 64

Colmenares, Margarita (1957-   ), 4-7
    awards and honors, 5
    B.S., civil engineering (Stanford University), 4
    career in civil engineering, 4-5
    chronology, 4
    community activities, 4-5
    professional organizations, 5
    White House Fellowship Program, 5
    work with U.S. Department of Education, 5

Conrad, Michael W., 61

D

Delgado, Jane (1953-   ), 8-11
    B.A., psychology (S.U.N.Y., New Paltz), 8
    chronology, 8
    educational career, 9
    health-care administration career, 9
    M.A., psychology (New York University), 9
    MA, urban policy and sciences (W. Averill
    Harriman School of Urban Policy Studies), 9
    Ph.D., clinical psychology
    (S.U.N.Y., Stony Brook), 9
    work in children's TV, 9

F

Faustina, Solis, 64

G

Garcia, Catalina Esperanza, 64-65

Gigli, Irma (1936-   ), 12-15
    career in immunology research, 12-13
    certification in dermatology (N.Y.U), 12